I0470179

FIREFIGHTER FATALITIES

IN THE

UNITED STATES

IN 2006

U.S. Department of Homeland Security

Federal Emergency Management Agency

U.S. Fire Administration

July 2007

In memory of all firefighters

who answered their last call in 2006

To their families and friends

To their service and sacrifice

U.S. Fire Administration

Mission Statement

As an entity of the Federal Emergency Management Agency (FEMA), the mission of the U.S. Fire Administration (USFA) is to reduce life and economic losses due to fire and related emergencies, through leadership, advocacy, coordination, and support. We serve the Nation independently, in coordination with other Federal agencies, and in partnership with fire protection and emergency service communities. With a commitment to excellence, we provide public education, training, technology, and data initiatives.

ACKNOWLEDGMENTS

This study of firefighter fatalities would not have been possible without the cooperation and assistance of many members of the fire service across the United States. Members of individual fire departments, chief fire officers, wildland fire service organizations such as the United States Forest Service (USFS), the National Park Service (NPS), the Bureau of Land Management (BLM), the Bureau of Indian Affairs (BIA), the United States Fish and Wildlife Service (FWS), as well as the United States Department of Justice (DOJ), the National Fire Protection Association (NFPA), the National Fallen Firefighters Foundation (NFFF), and many others contributed important information for this report.

C² Technologies, Inc., of Vienna, Virginia, conducted this analysis for the United States Fire Administration (USFA) under contract EME-2003-CO-0282.

The ultimate objective of this effort is to reduce the number of firefighter deaths through an increased awareness and understanding of their causes and how they can be prevented. Firefighting, rescue, and other types of emergency operations are essential activities in an inherently dangerous profession, and unfortunate tragedies do occur. This is the risk all firefighters accept every time they respond to an emergency incident. However, the risk can be reduced greatly through efforts to improve training, emergency scene operations, and firefighter health and safety initiatives.

Cover Photo: The funeral of Firefighter Hector "Sandy" McClune. Firefighter McClune suffered a heart attack on the scene of a wildland fire. Photo courtesy of Kathy McLaughlin/Poughkeepsie Journal.

FIREFIGHTER LIFE SAFETY INITIATIVES

(www.EveryoneGoesHome.com)

1. Define and advocate the need for a cultural change within the fire service relating to safety; incorporating leadership, management, supervision, accountability, and personal responsibility.

2. Enhance personal and organizational accountability for health and safety throughout the fire service.

3. Focus greater attention on the integration of risk management with incident management at all levels, including strategic, tactical, and planning responsibilities.

4. All firefighters must be empowered to stop unsafe practices.

5. Develop and implement national standards for training, qualifications, and certification (including regular recertification) that are equally applicable to all firefighters based on the duties they are expected to perform.

6. Develop and implement national medical and physical fitness standards that are equally applicable to all firefighters, based on the duties they are expected to perform.

7. Create a national research agenda and data collection system that relates to the initiatives.

8. Use available technology wherever it can produce higher levels of health and safety.

9. Thoroughly investigate all firefighter fatalities, injuries, and near misses.

10. Grant programs should support the implementation of safe practices and/or mandate safe practices as an eligibility requirement.

11. National standards for emergency response policies and procedures should be developed and championed.

12. National protocols for response to violent incidents should be developed and championed.

13. Firefighters and their families must have access to counseling and psychological support.

14. Public education must receive more resources and be championed as a critical fire and life safety program.

15. Advocacy must be strengthened for the enforcement of codes and the installation of home fire sprinklers.

16. Safety must be a primary consideration in the design of apparatus and equipment.

Sprinklers Save Firefighters' Lives Too.

Firefighter Life Safety Initiative No.15:

Strengthen advocacy for the enforcement of codes and the installation of home fire sprinklers.

TABLE OF CONTENTS

continued on next page

BACKGROUND

For three decades, the United States Fire Administration (USFA) has tracked the number of firefighter fatalities and conducted an annual analysis. Through the collection of information on the causes of firefighter deaths, the USFA is able to focus on specific problems and direct efforts toward finding solutions to reduce the number of firefighter fatalities in the future. This information also is used to measure the effectiveness of current programs directed toward firefighter health and safety.

Several programs have been funded by the USFA in response to this detailing of firefighter fatalities. For example, the USFA has sponsored significant work in the areas of vehicle operation safety and roadside incident safety. The data developed for this report are also used widely in other firefighter fatality prevention efforts, research, academia, and by the press.

In addition to the analysis, the USFA provides the list of onduty firefighter fatalities to the National Fallen Firefighters Foundation (NFFF). The Foundation tracks line-of-duty fatalities and applies their criteria (see Appendix D) to identify those firefighters to be honored at the events of the annual National Fallen Firefighters Memorial Weekend. Where the criteria are met, the fallen firefighter's next of kin, as well as members of the individual's fire department, are invited to the annual National Fallen Firefighters Memorial Service. The service is held at the National Emergency Training Center (NETC) in Emmitsburg, Maryland, during Fire Prevention Week in October of each year. Additional information regarding the Memorial Service can be found at www.firehero.org or by calling the NFFF at (301) 447-1365.

Other resources and information regarding firefighter fatalities, including current fatality notices, the National Fallen Firefighters Memorial database, and links to the Public Safety Officers' Benefit (PSOB) Program can be found at www.usfa.dhs.gov/fireservice/fatalities/

INTRODUCTION

This report continues a series of annual studies by the USFA of onduty firefighter fatalities in the United States. The USFA is the single public agency source of information for all onduty firefighter fatalities in the United States each year. This information is in the public domain and may be accessed on the USFA Web site at *www.usfa.dhs.gov/fireservice/fatalities/*.

The unique and specific objective of this study is to identify all onduty firefighter fatalities that occurred in the United States and its protectorates in 2006, and to present in summary form the circumstances surrounding each occurrence. The study is intended to help identify approaches that could reduce the number of firefighter deaths in future years.

In addition to the 2006 overall findings, this study includes information on the hazards to firefighters presented by engineered lumber when it is exposed to fire conditions.

The National Institute for Occupational Safety and Health (NIOSH) Fire Fighter Fatality Investigation and Prevention Program (*www.cdc.gov/niosh/fire/*) is also a unique and critically valuable counterpart to the comprehensive annual USFA study, reporting on indepth investigative findings for individual firefighter fatality incidents.

The National Fallen Firefighter Foundation maintains the list of firefighters who die in the line-of-duty and are honored during the annual National Fallen Firefighters Memorial Weekend. The names of these firefighters are commemorated on a plaque that is permanently installed at the National Fallen Firefighter Memorial in Emmitsburg, Maryland. A list of the firefighters being honored for 2006 (a total of 91) can be found in Appendix D.

WHO IS A FIREFIGHTER?

For the purpose of this study, the term "firefighter" covers all members of organized fire departments in all 50 States, the District of Columbia, and the Territories of Puerto Rico, the Virgin Islands, American Samoa, the Commonwealth of the Northern Mariana Islands, and Guam. It includes career and volunteer firefighters; full-time public safety officers acting as firefighters; State, Territory, and Federal government fire service personnel, including wildland firefighters; and privately employed firefighters, including employees of contract fire departments and trained members of industrial fire brigades, whether full- or part-time. It also includes contract personnel working as firefighters or assigned to work in direct support of fire service organizations.

Under this definition, the study includes not only local and municipal firefighters but also seasonal and full-time employees of the USFS, the BLM, the BIA, the FWS, the NPS, and State wildland agencies. The definition also includes prison inmates serving on firefighting crews; firefighters employed by other governmental agencies, such as the United States Department of Energy (DOE); military personnel performing assigned fire suppression activities; and civilian firefighters working at military installations.

WHAT CONSTITUTES AN ONDUTY FATALITY?

Onduty fatalities include any injury or illness sustained while on duty that proves fatal. The term "on duty" refers to being involved in operations at the scene of an emergency, whether it is a fire or nonfire incident; responding to or returning from an incident; performing other officially assigned duties such as training, maintenance, public education, inspection, investigations, court testimony, and fundraising; and being on call, under orders, or on standby duty except at the individual's home or place of business. An individual who experiences a heart attack or other fatal injury at home while he or she prepares to respond to an emergency is considered on duty when the response begins. A firefighter that becomes ill while performing fire department duties and suffers a heart attack shortly after arriving home or at another location may be considered on duty, since the inception of the heart attack occurred while the firefighter was on duty.

On December 15, 2003, the President of the United States signed into law the Hometown Heroes Survivors Benefit Act of 2003. After being signed by the President, the Act became Public Law 108-182. The law presumes that a heart attack or stroke are in the line of duty if the firefighter was engaged in nonroutine stressful or strenuous physical activity while on duty and he or she becomes ill while on duty or within 24 hours after engaging in such activity. The full text of the law is available in Appendix C on page 100 of this report.

The inclusion criteria for this study have been affected by this change in the law. Previous to December 15, 2003, firefighters who became ill as the result of a heart attack or stroke after going off duty needed to register some complaint of not feeling well while still on duty in order to be included in this study. For firefighter fatalities after December 15, 2003, firefighters will be included in this study if they become ill as the result of a heart attack or stroke within 24 hours of a training activity or emergency response. Firefighters who become ill after going off duty where the activities while on duty were limited to non-stressful tasks that did not involve physical exertion, such as clerical or administrative, or were nonmanual tasks, will not be included in this study.

A fatality may be caused directly by an accidental or intentional injury in either emergency or nonemergency circumstances, or it may be attributed to an occupationally related fatal illness. A common example of a fatal illness incurred on duty is a heart attack. Fatalities attributed to occupational illnesses also would include a communicable disease contracted while on duty that proved fatal, when the disease could be attributed to a documented occupational exposure.

Firefighter fatalities are included in this report even when death is considerably delayed after the original incident. When the incident and the death occur in different years, the analysis counts the fatality as having occurred in the year in which the incident took place. Three firefighters died in 2006 as a result of incidents that occurred in previous years. Two of the firefighters were injured as they fought structural fires and remained comatose for over a decade until the time of their deaths. The third firefighter was severely injured in a vehicle-related incident and never fully recovered. Information about these deaths is included in the appendix of this report, but they are not addressed in the body of the report unless the death affects retrospective statistical comparisons.

There is no established mechanism for identifying fatalities that result from illnesses such as cancer that develop over long periods of time and which may be related to occupational exposure to hazardous materials or toxic products of combustion. It has proved to be very difficult over the years to provide a complete evaluation of an occupational illness as a causal factor in firefighter deaths due to the following limitations: the exposure of firefighters to toxic hazards is not sufficiently tracked; the often delayed long-term effects of such toxic hazard exposures; and the exposures firefighters may receive while off duty.

SOURCES OF INITIAL NOTIFICATION

As an integral part of its ongoing program to collect and analyze fire data, USFA solicits information on firefighter fatalities directly from the fire service and from a wide range of other sources. These sources include the PSOB Program administered by the DOJ, the NIOSH, the Occupational Safety and Health Administration (OSHA), the Department of Defense (DOD), the National Interagency Fire Center (NIFC), and other Federal agencies.

The USFA receives notification of some deaths directly from fire departments, as well as from such fire service organizations as the International Association of Fire Chiefs (IAFC), the International Association of Fire Fighters (IAFF), NFPA, the National Volunteer Fire Council (NVFC), State fire marshals, State fire training organizations, other State and local organizations, fire service Internet sites, news services, and fire service publications. The USFA also keeps track of fatal fire incidents as part of its Major Fires Investigation Program and performs an ongoing analysis of firefighter injury and fatality data from the National Fire Incident Reporting System (NFIRS).

Figure 1. Onduty Firefighter Fatalities (1977-2006)

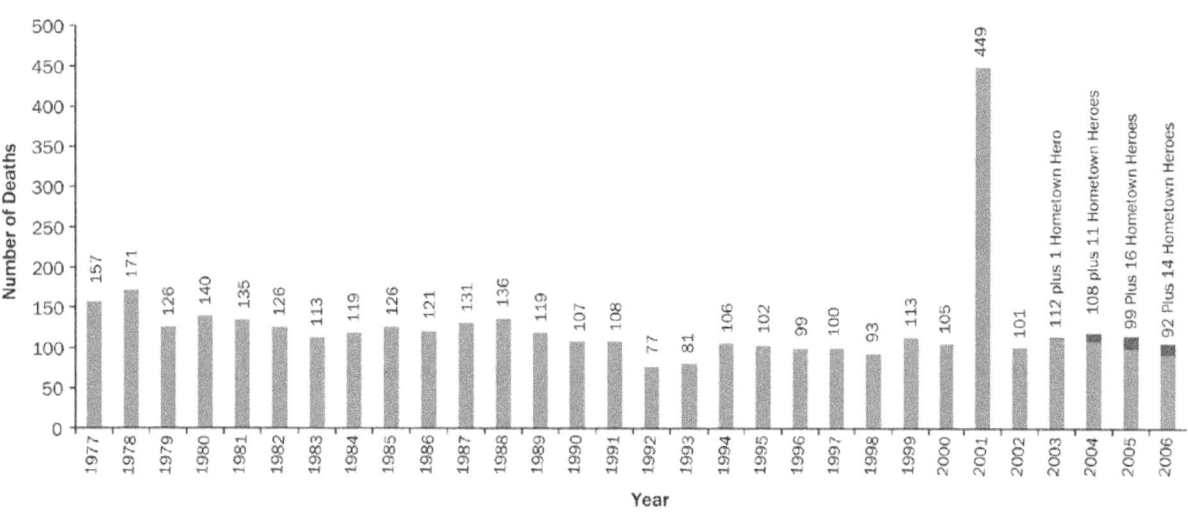

In 2001, the largest loss of firefighter's lives in a single incident occurred as a result of the attacks on the World Trade Center (WTC) in New York City on September 11th. A total of 344 firefighters were killed in the attacks and resulting collapses. When conducting multiyear comparisons of firefighter fatalities in this report, it may be necessary to set these deaths apart for illustrative purposes. This action is by no means a minimization of the supreme sacrifice made by these firefighters.

CAREER AND VOLUNTEER DEATHS

Firefighter fatalities in 2006 included 77 volunteer firefighters and 29 career firefighters (Figure 2). Among the volunteer firefighter fatalities, 61 were from local or municipal volunteer fire departments, and 16 were part-time or full-time members of wildland fire agencies. One career firefighter was a member of an industrial fire department and the rest were members of local or municipal fire departments. Six of the firefighters who died in 2006 were female and 100 were male. In the last decade, female firefighter deaths have ranged from none in 1998 to 6 in 2004 and in 2006.

The six female firefighter deaths in 2006 represent 5.7 percent of the total number of firefighter deaths. This is the highest percentage of female deaths in over a decade.

Figure 2. Career and Volunteer Firefighter Deaths In 2006

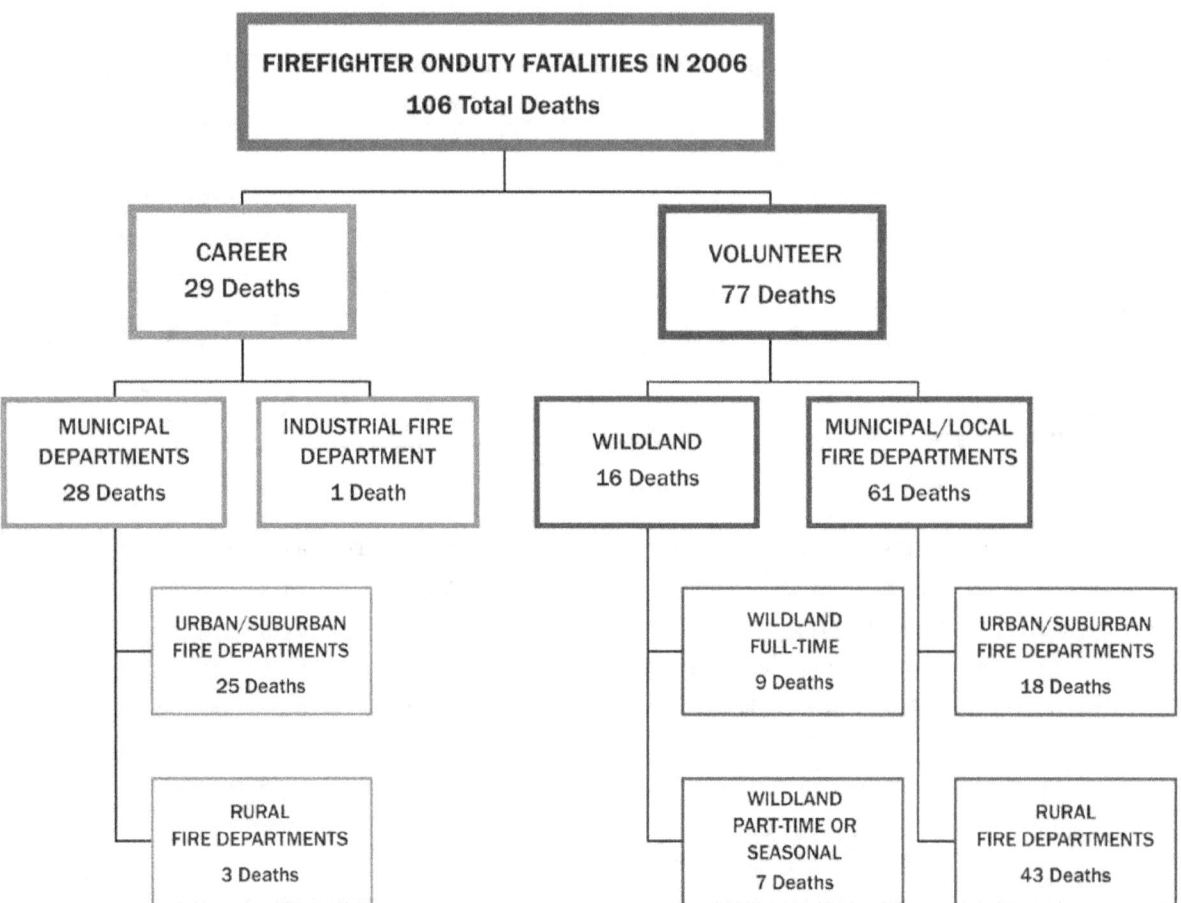

MULTIPLE FIREFIGHTER FATALITY INCIDENTS

The 106 deaths in 2006 resulted from a total of 95 fatal incidents. There were 6 firefighter fatality incidents where 2 or more firefighters were killed in 2006, claiming a total of 17 firefighters' lives.

Table 1. Multiple Firefighter Fatality Incidents

Year	Number of Incidents	Total Number of Deaths
2006	6	17
2005	4	10
2004	3	6
2003	7	20
2002	9	25
2001	8	362
2001 w/o WTC	7	18
2000	5	10
1999	6	22
1998	10	22
1997	8	17

- In February, two Alabama volunteer firefighters were killed when they were crushed by a collapsing wall at a commercial structure fire. Approximately 4 hours into the incident, firefighters stretched a hoseline to the front of the structure to extinguish hot spots. The front wall of the structure collapsed outward, and both firefighters were trapped in the debris.

- On August 4th, two wildland contract pilots were killed in the crash of their helicopter in California. The aircraft suffered a tail rotor failure during a water dip and crashed, killing both firefighters.

- On August 13th, four wildland firefighters were killed in the crash of a helicopter in Idaho. For reasons unknown, the helicopter crashed during a crew transport mission, and all occupants were killed.

- On August 27th, two New York City career firefighters were killed as the result of a structural collapse during a fire in a commercial occupancy. Both firefighters were trapped in the basement of the structure and their removal involved a major rescue effort.

- On September 6th, a career wildland firefighter and a wildland contract pilot were killed in the crash of their observer aircraft. For reasons unknown, the aircraft hit trees, broke apart, and burned.

- On October 26th, five California wildland firefighters were killed when their structural defense position was overcome by rapidly advancing flames. Three firefighters were killed at the scene, one firefighter died enroute to the hospital, and one firefighter died 5 days after the event.

The Atlanta Fire Department in Georgia suffered the loss of two firefighters in 2006 as the result of unrelated incidents. In April, an onduty firefighter suffered a heart attack and later died. A residential structure fire in November claimed the life of another Atlanta firefighter.

WILDLAND FIREFIGHTING DEATHS

In 2006, 22 firefighters were killed during activities involving brush, grass, or wildland fire fighting. This total includes part-time and seasonal wildland firefighters, full-time wildland firefighters, and municipal or volunteer firefighters whose deaths are related to a wildland fire (Figure 3).

Aircraft-related incidents claimed the lives of eight wildland firefighters in 2006. This is the largest number of firefighters to be killed in aircraft-related incidents in over a decade. These firefighters were killed in three incidents: one incident involved a fixed-wing aircraft and the other

Photo courtesy of The Denver Post.

The funeral of Denver Fire Lieutenant Richard Montoya. Lieutenant Montoya died fighting a residential structure fire.

two incidents involved helicopters. These incidents are described in more detail in the multiple firefighter fatality section of this report above.

The single incident that took the largest number of firefighters' lives in 2006 was also wildland-related. In October, five California wildland firefighters were killed when their defensive position was overrun by rapid fire progress.

Two firefighters were killed in 2006 as the result of incidents at prescribed burns in April. A Tennessee forestry technician was killed when he was struck by a falling tree at a prescribed burn; an Oklahoma volunteer firefighter was killed when he was run over by the fire apparatus he was operating at a prescribed burn.

In addition to the California multiple firefighter fatality incident, two firefighters were killed when they were overrun by fire progress at separate incidents. An Oklahoma firefighter was killed in March when his wildland apparatus was overcome by advancing flames; the firefighter dismounted his apparatus to assist another firefighter, and he was severely burned. A Utah wildland firefighter was killed when he was overrun by rapidly advancing fire conditions at a wildland fire incident. The Utah firefighter was scouting a fire as a part of the incident command team when fire unexpectedly advanced toward his position. He deployed his fire shelter but the intensity of the fire was too extreme and he was killed.

Three firefighters died of heart attacks associated with wildland incidents. A New York volunteer firefighter suffered a heart attack in November at the scene of a wildland fire on the grounds of a local school. Two firefighters suffered heart attacks after the completion of firefighting duties at wildland fires in Kansas and North Carolina.

A Texas firefighter was killed in a tanker (tender) rollover at a wildland fire. The firefighter had been on the scene of the wildland incident for over 7 hours and was the driver of a military 6X6 that had been converted for firefighting duties. The apparatus lost traction in soft soil and rolled over. The firefighter was ejected from the vehicle and killed.

A South Dakota firefighter was killed when he was struck by a recoiling rope and attached hardware. The apparatus being operated by the firefighter had become stuck in a field, and was being pulled free. The tow rope failed, and the firefighter was struck by hardware at the end of the tow rope as it recoiled from the failure. The firefighter was seated at the wheel of the apparatus and was struck as the object came through the windshield.

Figure 3. Firefighter Fatalities Related to Wildland Firefighting (1997-2006)

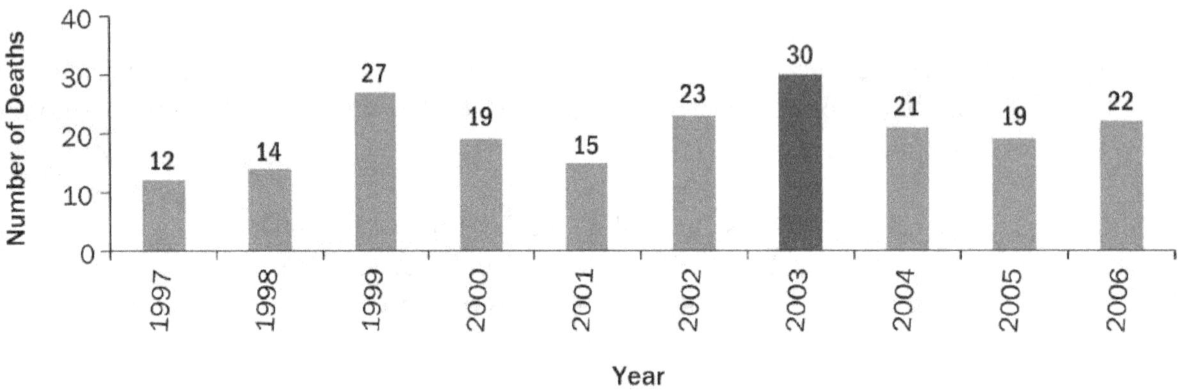

Table 2. Firefighter Deaths Associated with Wildland Firefighting

Year	Total Number of Deaths	Number of Fatal Incidents	Number of Firefighters Killed in Multiple-Death Incidents
2006	22	13	13
2005	19	15	6
2004	21	21	0
2003	29	21	10
2002	23	14	13
2001	15	9	9
2000	17	14	6
1999	26	25	2
1998	14	13	2
1997	12	10	4

Table 3. Wildland Fire Fighting Aircraft Deaths

Year	Total Number of Deaths	Number of Fatal Incidents
2006	8	3
2005	6	2
2004	3	3
2003	7	4
2002	6	3
2001	6	3
2000	6	5
1999	0	0
1998	3	2
1997	5	3

Figure 4. Firefighter Fatalities per 100,000 Fires (1996-2005)

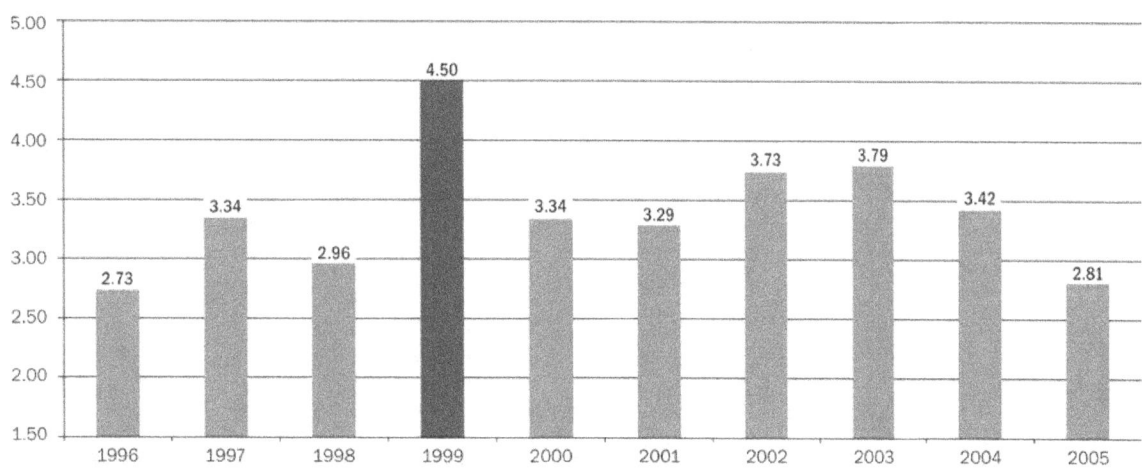

TYPE OF DUTY

Activities related to emergency incidents resulted in the deaths of 61 firefighters in 2006 (Figure 5). This includes all firefighters who died while responding to an emergency, while at an emergency scene, while returning from the emergency incident, and other emergency-related activities. Nonemergency activities accounted for 45 fatalities. Nonemergency duties include training, administrative activities, performing other functions that are not related to an emergency incident, and postincident fatalities where the firefighter does not experience the illness or injury during the emergency.

A multiyear historical perspective concerning the percentage of firefighter deaths that occurred during emergency duty is presented in Table 4. The data for 2003 to 2006 is skewed somewhat by the inclusion of firefighters covered by the changes resulting from the Hometown Heroes Act of 2003.

Figure 5. Firefighter Deaths by Type of Duty (2006)

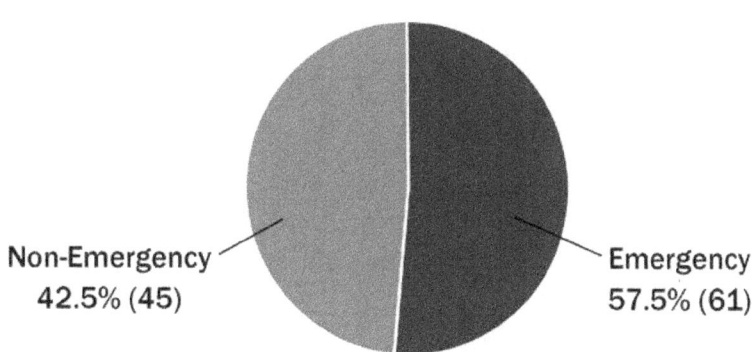

Non-Emergency
42.5% (45)

Emergency
57.5% (61)

Table 4 – Emergency Duty Firefighter Deaths

Year	Percentage of All Deaths	Percentage of All Deaths without Hometown Heroes
2006	57.5	66.3
2005	52.1	60.6
2004	68.9	75.9
2003	69	69.6
2002	73	N/A
2001	65	N/A
2001 with WTC	92	N/A
2000	71	N/A
1999	87	N/A
1998	77	N/A
1997	81	N/A

The number of deaths by type of duty being performed in 2006 is shown in Table 5 and presented graphically in Figure 6. As has been the case for most years, fireground duties are the most common type of duty for firefighters killed while on duty.

Table 5 – 2006 Firefighter Deaths by Type of Duty

Type of Duty	Number of Deaths
Fireground Operations	36
Other On Duty	21
Responding/Returning	15
Training	9
Nonfire Emergencies	5
After an Incident	20
Total	**106**

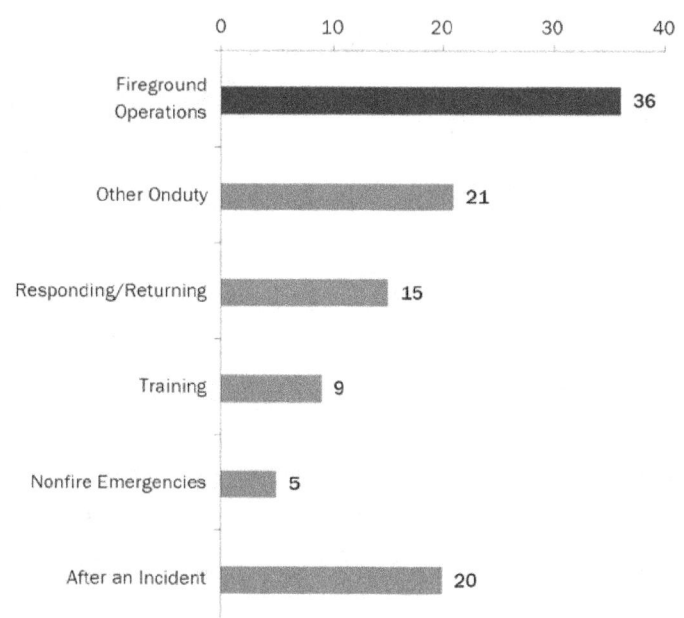

Figure 6. Fatalities by Type of Duty (2006)

Fireground Operations

The most hazardous duty for firefighters in 2006, as in most years, was working on the scene of a fire incident. Thirty-six firefighters died while engaging in activities at the scene of a fire in 2006:

- Five California firefighters were killed when their position was overrun by fire during a wildland fire incident in October.
- Two Alabama firefighters were killed when they were crushed by a collapsing wall at the scene of a commercial structure fire in February.
- Two wildland firefighters were killed in the crash of their helicopter during a fire in California in August.
- Two New York City firefighters were killed when they were caught in the collapse of a building during a fire fight in August.
- Two California wildland firefighters were killed in a plane crash as they worked on the scene of a wildland fire in September.
- In addition to the multiple firefighter fatality incidents described above, 11 firefighters suffered fatal traumatic injuries at structure fires in 2006:
  - A New Jersey volunteer firefighter died of smoke inhalation at a fire in his home. After discovering the fire, the firefighter evacuated others and fought the fire as other firefighters responded.

Two firefighters died in 2006 when fire-weakened floors gave way and dropped the firefighters into the basement of the structure. Both of these incidents involved the failure of engineered lumber products under fire conditions.

Photo by Paris L. Gray, Courier-Post Newspaper, Cherry Hill, New Jersey.

The funeral of Firefighter Edward Marbet. Firefighter Marbet died fighting a structure fire in his home.

- A Mississippi inmate volunteer firefighter became disoriented in a residential structure fire.
- A New Jersey volunteer firefighter died during a rescue attempt in a residential structure fire. The firefighter had located a civilian fire victim and was crawling toward the exit when a floor collapse claimed both lives.
- A Colorado career fire officer became disoriented or trapped in a residential structure fire. He died 7 days later.
- An Indiana volunteer firefighter died after falling into the basement of a residential structure fire when fire-weakened flooring gave way.
- A Wisconsin firefighter fell through the fire-weakened floor of a residential structure and was trapped in the basement.
- A New Jersey firefighter was trapped by rapid fire progress in an apartment fire.
- A Maryland firefighter was trapped within a residential structure as he and other firefighters attempted to exit the building due to rapid fire development.
- An Indiana firefighter became disoriented in a large residence and died of smoke inhalation and burns.
- A Georgia firefighter became disoriented during a fire in an abandoned residence. He suffered fatal burns and died 6 days after the fire.
- A Texas firefighter was killed as he fought a structural fire in a commercial building. A collapse occurred and buried the firefighter under debris.
- Eight firefighters suffered heart attacks at fire scenes in 2006:

- Seven of the heart attacks occurred at the scene of structure fires. At least two of the fatal incidents occurred during mutual-aid responses.
- One heart attack occurred at a wildland fire.
- Two firefighters were overrun by fire progress at wildland fires. The deaths occurred in Oklahoma and Utah.
- A Texas firefighter was killed when the water tender he was operating rolled over during a wildland fire.
- A South Dakota firefighter was killed when he was struck by hardware attached to a tow rope. When the rope attachment failed during an attempt to pull an apparatus from a field, the firefighter was struck by the recoiling rope and attachment.

Other On Duty

In 2006, 21 firefighters died while on duty engaged in activities that were not associated with the response to any particular emergency:

- Four firefighters were killed in the crash of a helicopter in Idaho. The firefighters were in the middle of a crew replacement when the crash occurred.
- A total of nine firefighters suffered onduty heart attacks that were not associated with training or incident response activities:
  - Three firefighters suffered heart attacks while on duty in the fire station. These deaths occurred in South Carolina, Connecticut, and New Jersey.
  - Two firefighters died of heart attacks after participating in fire department parades. Both incidents occurred in New York.
  - A North Carolina firefighter died of a heart attack that struck as he purchased parts for a forestry bulldozer.
  - A North Carolina firefighter suffered a heart attack during a standby at a public fireworks event.
  - An Ohio career firefighter suffered a heart attack while at home for lunch.
  - A New Hampshire firefighter had a heart attack while transporting a fire truck to another town for maintenance.
- Two firefighters were killed as they were engaged in prescribed burns. A Tennessee firefighter was struck by a falling tree and an Oklahoma firefighter was killed when he was run over by the apparatus he was driving.
- Two firefighters were killed in crashes that involved vehicles while they were on duty but not assigned to an incident or training activity:
  - A Kansas firefighter was killed as he drove his personal vehicle to retrieve surplus equipment from the state forestry department. For unknown reasons, his car crossed the center line and collided with an oncoming truck.
  - A Kentucky firefighter was involved in a crash while operating his personal vehicle as he traveled back to the fire station to help with a fundraising event.

- Two firefighters were killed as the result of being struck by vehicles while on duty:
  - A North Carolina firefighter was killed while he was engaged in painting hydrant markings in the roadway. The fire apparatus he was driving was positioned to shield his work area but the apparatus was propelled forward and struck the firefighter after a vehicle crashed into the apparatus.
  - A Pennsylvania firefighter was struck and killed by a pickup truck in a bank parking lot as he carried the proceeds of his department's fundraising activities to the bank.
- A New York firefighter was electrocuted and died while retrieving a tarp from a fire-damaged business. The firefighter came into contact with an electrified sign that was not properly wired.
- A Pennsylvania firefighter suffered an aortic aneurysm while driving to the fire station to perform his assigned maintenance duties.

Responding/Returning

Fifteen firefighters died while responding to or returning from emergency incidents in 2006. Fourteen firefighters died while responding to an emergency incident, and one died while returning from an emergency.

- Six firefighters were killed in vehicle crashes while responding:
  - A New Mexico firefighter was involved in a crash while responding to an emergency medical services (EMS) incident. His vehicle entered a curve, left the roadway, and crashed into a utility pole. He was not wearing a seatbelt.
  - A North Carolina firefighter responding in his personal vehicle to a smoke investigation was killed when he struck a tractor-trailer that was blocking the roadway. Fog may have contributed to the crash.
  - An Alabama firefighter was killed when the water tender in which she was a passenger rolled over the guard rail on a bridge and fell into the creek bed below.
  - An Ohio firefighter was killed when her personal vehicle entered a curve, left the roadway, and crashed. She was responding to an EMS incident.
  - An Indiana chief officer was killed when the water tender he was driving rolled over multiple times. The chief officer was ejected during one of the rolls.
  - A Minnesota firefighter was killed during a response when he failed to stop at a stop sign and was struck by a vehicle at an intersection.
- Seven firefighters died of heart attacks that struck while the firefighter was responding to an emergency:
  - A New York firefighter collapsed outside his fire station as he arrived for a response. Firefighters discovered him when the incident was concluded.
  - A New York fire police officer began to feel ill during his response to a hazardous materials incident. When he arrived on scene, he was treated by other firefighters.
  - An Illinois chief officer began feeling ill during a response. He told other firefighters to go to the incident and went home. He suffered a heart attack after arriving home.

Photo courtesy of the Posey Township Volunteer Fire Department.

Assistant Fire Chief Errett Miller was killed in the crash of a water tender while responding to an arson-caused mutual aid structure fire.

- – A Mississippi firefighter was killed when he suffered a heart attack and fell from the bed of a pickup truck while responding.
- – A New York fire police officer became ill and later died of a heart attack as he prepared to respond from his fire station.
- – A North Carolina firefighter died after suffering a heart attack while driving a heavy rescue apparatus to an incident. The apparatus left the roadway and crashed.
- – A Maryland firefighter suffered a heart attack while responding in his personal vehicle.
- A North Carolina firefighter was killed as he prepared to respond to an EMS incident during severe weather. A tornado struck his home as he departed and he was killed.
- A Tennessee firefighter was killed as he drove a fire department water tender back from an incident response. The vehicle's brakes failed and the water tender was involved in a crash.

Table 6. Firefighter Deaths While Responding to or Returning From an Incident

Year	Number of Firefighter Deaths
2006	15
2005	22
2004	23
2003	36
2002	13
2001	23
2000	19
1999	26
1998	14
1997	21

Training

In 2006, 9 firefighters died while they were engaged in training activities.

- Four firefighters died during physical fitness training:
    - A California firefighter suffered a heart attack after leading his crew on a conditioning hike.
    - A Georgia firefighter died of a heart attack after completing a treadmill workout.
    - A Nevada firefighter suffered a heart attack while exercising on a treadmill.
    - A North Carolina firefighter suffered a heart attack while running wind sprints outside his fire station.
- A Washington firefighter died of complications of a near drowning. He had been engaged in surf rescue training using personal watercraft when he was knocked from his watercraft by an unexpected wave.
- A Texas industrial firefighter suffered a fatal cerebrovascular accident (CVA) (stroke) while assisting with live-fire training exercises.
- A New York firefighter was killed as a heavy rescue apparatus was pulled out of the fire station during training. The firefighter was on top of the apparatus and was crushed between the apparatus and the door frame.
- A Virginia firefighter suffered a heart attack during a vehicle extrication demonstration.
- A West Virginia firefighter died of a CVA that he experienced during a break from training activities in his fire station.

Table 7 offers a multiyear perspective on training deaths.

Table 7 – Firefighter Deaths During Training

Year	Number of Firefighter Deaths
2006	9
2005	14
2004	13
2003	12
2002	11
2001	14
2000	13
1999	3
1998	12
1997	5

Nonfire Emergencies

Five firefighters died when they became ill or were injured while on the scene of emergencies that did not involve fire.

- Heart attacks struck three firefighters as they worked on the scene of motor vehicle incidents:
  - A Missouri fire captain collapsed after assisting with cardiopulmonary resuscitation (CPR) and carrying the victim of a motor vehicle crash up an embankment to the roadway above.
  - A New York firefighter became ill on the scene of a motor vehicle crash after assisting with equipment movement at the incident.
  - An Ohio fire officer suffered a heart attack on the scene of a car/pedestrian incident.
- A Mississippi firefighter died from a blood infection after cutting his finger on the scene of a motor vehicle crash. He died seven days after the incident.
- A Michigan firefighter was killed when she was struck by a vehicle on the scene of a weather-related crash involving multiple vehicles. An oncoming car slid on icy roads and struck the firefighter.

After the Incident

Twenty firefighters died after the conclusion of their onduty activity.

- Fourteen firefighters died after going off duty without complaining of illness while they were on duty. Thirteen of these deaths were due to heart attacks and one was due to an epileptic condition.
- Five firefighters suffered heart attacks after the conclusion of their incident response but prior to going off duty.

- An Illinois firefighter suffered a stroke after returning to quarters after a response to an automatic fire alarm.

Career, Volunteer, and Wildland Deaths by Type of Duty

Figure 7 depicts career, volunteer, and wildland firefighter deaths by type of duty. Wildland career, wildland seasonal, and wildland contractor deaths were grouped together. As in past years, there were a disproportionate number of fatalities experienced by volunteer firefighters responding to and returning from alarms as compared to career firefighters. Fourteen volunteer firefighter deaths occurred while responding and 1 occurred while returning from an emergency. Of the responding deaths, six were due to vehicle crashes.

Figure 7. Career, Volunteer, and Wildland Deaths by Type of Duty (2006)

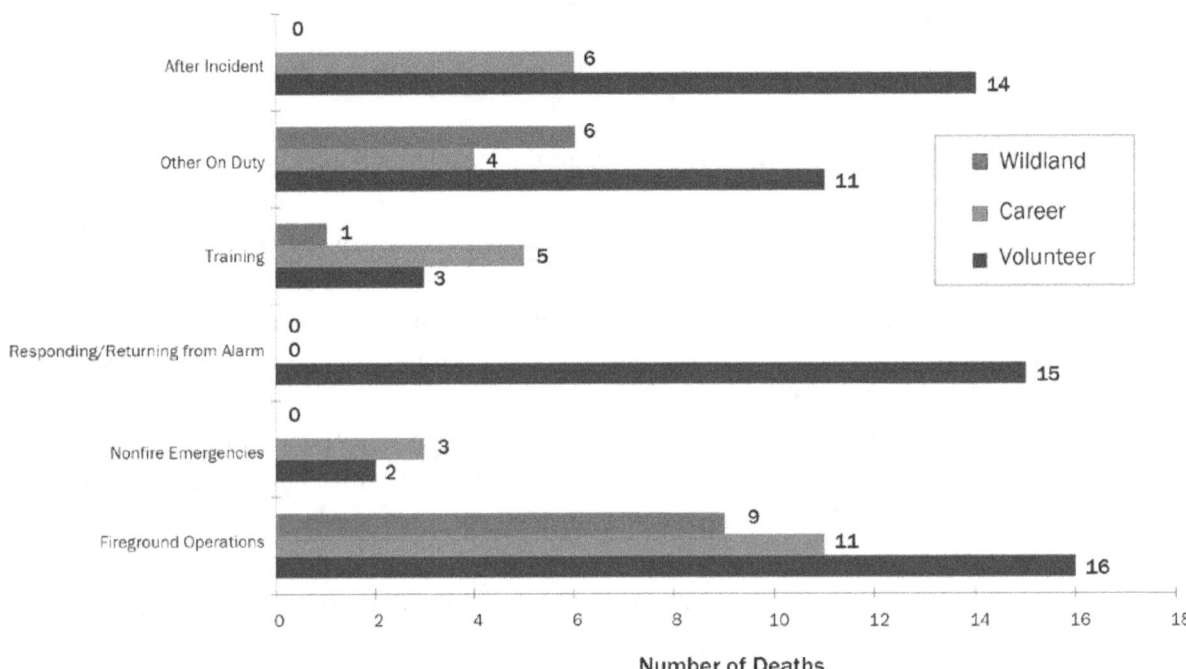

Type of Emergency Duty

In 2006, 55 firefighters died while engaged directly in the delivery of emergency services. This number includes deaths resulting from injuries sustained on the incident scene or enroute to the incident scene, and firefighters who became ill on an incident scene and later died. It does not include firefighters who became ill or died while returning from an incident (such as a vehicle collision while returning from an incident). Figure 8 shows the number of firefighters killed in firefighting, EMS, technical rescue-related incidents, and other emergency incidents in 2006.

Forty-four firefighters were killed during firefighting duties; 9 firefighters were killed on EMS calls; and 2 firefighters were killed at emergencies that involved hazardous materials and hazardous structural conditions.

Figure 8. Type of Emergency Duty (2006)

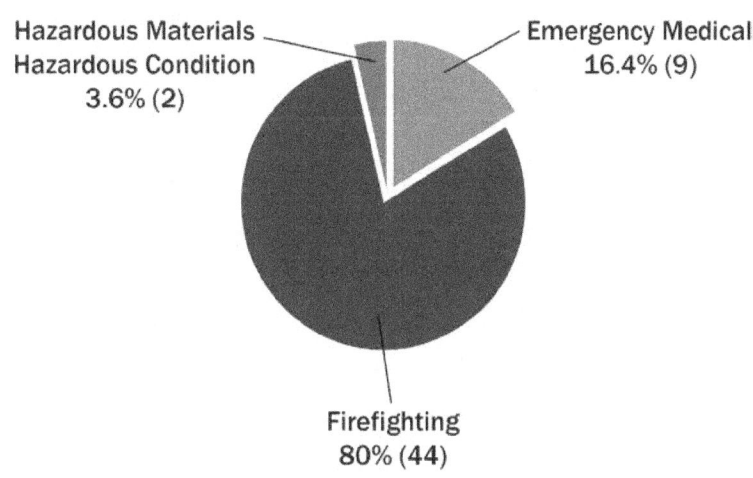

Note: 55 of 106 deaths occurred during emergency responses.

Photo courtesy of Tom Bushey and the Times Herald-Record.

Fire Police Officer David Smith died of a heart attack after responding to a hazardous materials incident.

CAUSE OF FATAL INJURY

The term "cause of injury" refers to the action, lack of action, or circumstances that resulted directly in the fatal injury. The term "nature of injury" refers to the medical cause of the fatal injury or illness; this is often referred to as the physiological cause of death. A fatal injury usually is the result of a chain of events; the first of which is recorded as the cause.

Table 8 and Figure 9 show the distribution of deaths by cause of fatal injury or illness.

Table 8 – Cause of Fatal Injury (2006)

Cause	Number
Stress/Overexertion	54
Vehicle Collision	19
Caught/Trapped	13
Collapse	8
Struck by	6
Lost	3
Contact/Exposure	2
Other	1
Total	**106**

Figure 9. Fatalities by Cause of Fatal Injury (2006)

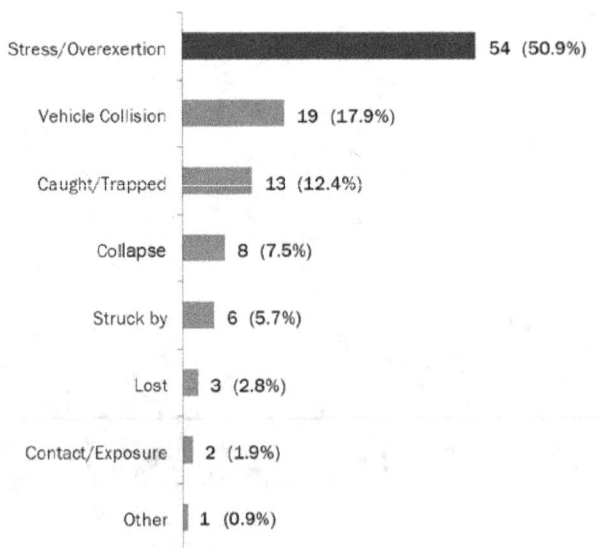

Stress/Overexertion 54 (50.9%)
Vehicle Collision 19 (17.9%)
Caught/Trapped 13 (12.4%)
Collapse 8 (7.5%)
Struck by 6 (5.7%)
Lost 3 (2.8%)
Contact/Exposure 2 (1.9%)
Other 1 (0.9%)

Stress or Overexertion

Stress or overexertion is a general category that includes all firefighter deaths that are cardiac or cerebrovascular in nature, such as heart attacks and strokes, and other events such as extreme climatic heat exposure. Classification of a firefighter fatality in this cause of fatal injury category does not indicate that a firefighter was in poor physical condition.

Firefighting is extremely strenuous physical work and is likely one of the most physically demanding activities that the human body performs.
- Fifty-four firefighters died in 2006 as a result of stress/overexertion:
  - Forty-nine of the stress deaths were heart attacks.
  - Three firefighter deaths were due to CVAs.
  - One firefighter fatality was due to an aortic aneurysm.
  - One firefighter fatality was epilepsy-related.
- If the Hometown Heroes deaths in 2006 are set aside for analysis purposes, 43.5 percent of firefighter fatalities in 2006 were caused by stress or overexertion.

Table 9 – Deaths Caused by Stress or Overexertion

Year	Number	Percent of Fatalities
2006	54	50.9
2005	62	53.9
2004	66	56.4
2003	51	45.9
2002	38	38.0
2001	43	40.9*
2000	46	44.6
1999	56	49.5
1998	43	46.2
1997	41	41.0

* Does not include the firefighter deaths of September 11, 2001, in New York City.

Information on chronic exposure to diesel exhaust for firefighters can be found at www.richter-foundation.org/home.html

Vehicle Crashes

After stress or overexertion, the perennial cause of fatal injury resulting in the most firefighter fatalities is vehicle crashes. In many cases, these deaths appear to have been preventable (Figure 10).

- Nineteen firefighters were killed in 2006 as a result of vehicle crashes.
- Eight of these deaths occurred in three aircraft crashes.
- Eleven firefighters were killed in nonaircraft vehicle crashes:
  - Five crashes involved water tenders.
  - Five crashes involved the firefighter's personal vehicle.
  - One crash involved a fire department pickup truck.
- No seatbelt was used in three of the four cases where seatbelts were available and the status of their use is known.

Water tenders are involved in a disproportionate number of fatal crashes. The USFA developed "Safe Operation of Fire Tankers." It is available for download at: www.usfa.dhs.gov/downloads/pdf/publications/fa-248.pdf

---

Figure 10. Firefighter Fatalities in Vehicle Collisions

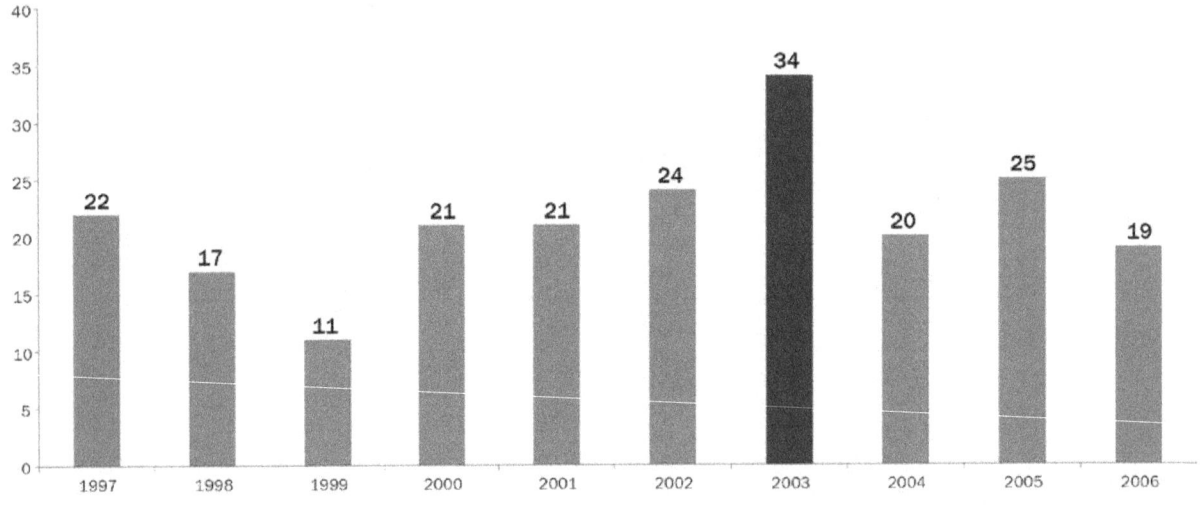

The 5 personal vehicle crash deaths in 2006 brings to 69 the number of firefighter deaths in personal vehicle crashes since 1990. Many of these deaths involved excessive speed and the lack of seatbelt use.

Caught or Trapped

Thirteen firefighters were killed when they were caught or trapped in 2006. This classification covers firefighters trapped in wildland and structural fires who were unable to escape due to rapid fire progression and its byproducts of smoke, heat, toxic gases, and flame. This classification also includes firefighters who are killed by drowning and those who were trapped and crushed.

- Five California wildland firefighters were killed when a fire progressed rapidly toward their position and they did not have time to escape.
- Three firefighters were trapped by advancing fire and smoke in residential structure fires in New Jersey, Maryland, and Georgia.
- A New Jersey firefighter was trapped by fire progression in a fire in his residence. The firefighter fought the fire while others responded but was overcome while trying to escape.
- An Oklahoma wildland firefighter was overrun by advancing fire as he operated a brush truck. The firefighter exited the apparatus to assist another firefighter and received significant burns.
- A wildland command officer in Utah was overrun by advancing fire. He deployed his fire shelter but was killed.
- A Washington firefighter died as the result of complications of a near-drowning that occurred during personal watercraft training.
- A New York firefighter was caught between the top of a heavy rescue apparatus and the station's door frame as the apparatus was pulled out of the station. The firefighter had been on top of the apparatus checking on supplies that were stored in rooftop compartments.

Collapse

Eight firefighters died in 2006 as the result of structural collapses. This was the fourth most common cause of fatal injury for firefighters in 2006.

- Two Alabama firefighters were killed when the front wall of a commercial structure fell and crushed them.
- Two New York City firefighters were killed when the floor of a commercial building failed during firefighting operations. Both firefighters were dropped into the basement and their removal by other firefighters was very complicated.
- Two firefighters were killed in separate incidents where engineered wood structural components failed under fire conditions. Both fires occurred in residences and both firefighters fell into the basement of the structure and were unable to escape. These incidents occurred in Indiana and Wisconsin.
- A New Jersey firefighter was killed when the floor of a residential structure collapsed under fire conditions and dropped him and the fire victim he was rescuing into the basement of the house.
- A Texas firefighter was killed when he was crushed by the structural failure of a wall at a commercial structure fire.

Photo courtesy of the Denison Fire Department.

Firefighter Philip Townsend was killed when he was crushed by a structural collapse at the scene of a fire in a commercial building.

Three firefighters were killed in two commercial building collapse incidents during fires in 2006. At both fires, firefighters were fighting the fire in a defensive strategy when the collapse occurred.

Struck by Object

Being struck by an object was the fifth leading cause of fatal firefighter injuries in 2006, as it was in 2005.

Six firefighters died after being struck by vehicles or other objects while on duty:

- A South Dakota firefighter was struck with hardware on the end of a tow rope when the rope failed. The rope recoiled and the hardware on the end traveled through the windshield of the apparatus, striking the firefighter in the head.
- A Michigan firefighter was struck by a vehicle as she worked on the scene of a motor vehicle crash in icy weather.
- A Tennessee firefighter was struck by a falling tree during a prescribed burn.
- A North Carolina firefighter was killed when the truck he was using to shield a work area on a local road was struck from behind. The apparatus was pushed forward, striking the firefighter.

- A Pennsylvania firefighter was struck by a car as he walked in a bank parking lot. The firefighter was delivering the proceeds of a fire department fundraiser to the bank.
- A North Carolina firefighter was struck by debris from a tornado as he began his response to an emergency.

Lost or Disoriented

Three firefighters died in 2006 when they became lost or disoriented in a structure during a fire. The incidents occurred in Mississippi, Colorado, and Indiana.

Enclosed Structure Firefighter Fatalities

A recent analysis of 444 firefighter fatalities that took place while the firefighter was on the scene of a structure fire has revealed the degree of danger associated with opened and enclosed structure fires. An opened structure is one that possesses numerous openings for firefighter access and egress. An enclosed structure is one with limited openings. Many "big box" retail stores incorporate the enclosed design.

The study also determined the degree of safety provided by the strategy and tactics used during these operations. The analysis considered structural fire firefighter fatalities from January 1, 1990, through December 31, 2006. The study found that 187 or 84 percent of the firefighter fatalities occurred in an enclosed structure while 36 or 16 percent of the fatalities occurred in an opened structure.

In all cases, an aggressive interior attack was used. Also, of the structure fires resulting in multiple firefighter fatalities, 34 or 87 percent occurred in an enclosed structure while 5 or 13 percent occurred in an opened structure.

The analysis concluded that over a 16-year time span, firefighters using an aggressive interior attack in enclosed structures died far more often, in greater numbers, and with greater multiple line-of-duty deaths than those using the same tactical approach in opened structure fires. In response to these significant findings, it is important that departments act to prevent additional firefighter deaths by adopting and implementing more appropriate enclosed structure tactics and standard operating guidelines (SOGs) for use during extremely dangerous enclosed structure fires.

More information on this work, contact Captain William Mora at capmora@aol.com Check *www.usfa.dhs.gov/fireservice/fatalities/* in coming weeks for a USFA Technical Report Series study on this topic.

Data source: USFA - National Fire Data Center

Contact/Exposure

Two firefighters were killed in 2006 when they came into contact with or were exposed to harm:

- A New York firefighter was electrocuted as he retrieved a fire department salvage tarp from a local restaurant that had experienced a recent fire. The firefighter came into contact with an electrified sign that was not properly grounded.

- A Mississippi firefighter died as the result of a blood infection contacted through a cut finger at a motor vehicle crash.

Other

One firefighter died in 2006 of a cause that is not categorized above. A Virginia volunteer firefighter died during an extrication demonstration of dilated cardiomyopathy.

Photo courtesy of the Denver Fire Department.

Denver Fire Department Lieutenant Richard Montoya died fighting a residential structure fire. This photo-graph shows the stairway firefighters ascended to attack the fire.

NATURE OF FATAL INJURY

Table 10 and Figure 11 show the distribution of the 106 firefighter deaths that occurred in 2006 by the medical nature of the fatal injury or illness.

Table 10 – Nature of Fatal Injury (2006)

Nature	Number
Heart Attack	50
Internal Trauma	24
Asphyxiation	12
Burns	8
Crushed	5
CVA	4
Electrocution	1
Other	2
Total	**106**

Figure 11. Fatalities by Nature of Fatal Injury (2006)

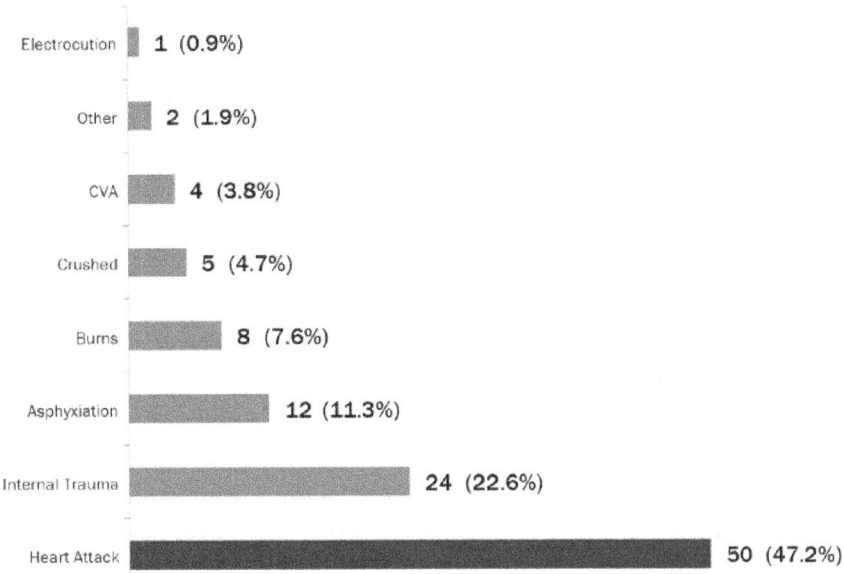

Heart Attack

Heart attacks were the most frequent cause of death for 2006, with 50 firefighter deaths, down from 55 heart attack deaths in 2005, and 61 heart attack deaths in 2004. Figure 12 provides a detailed breakdown of heart attacks by type of duty.

Figure 12. Heart Attacks by Type of Duty (2006)

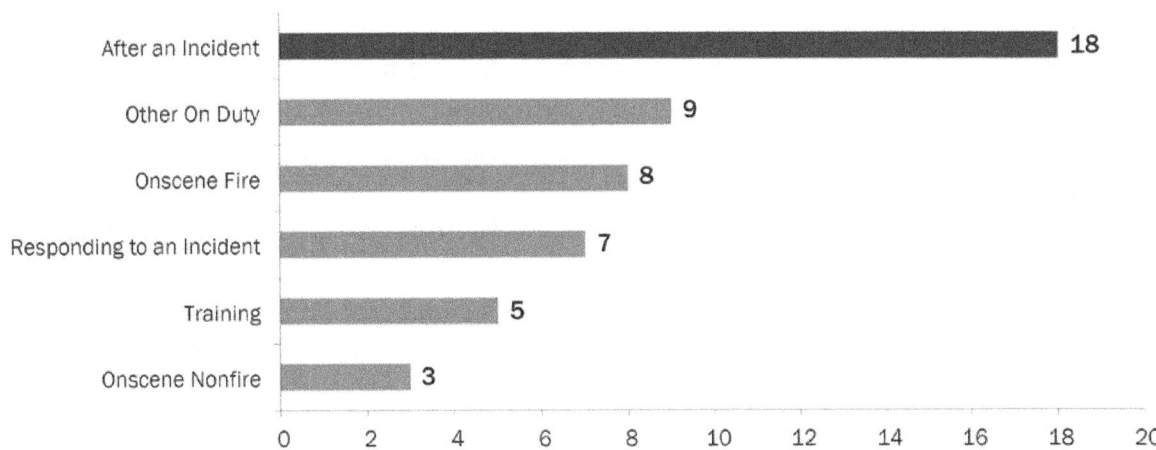

- Eighteen firefighters died of heart attacks that struck after the conclusion of an incident response or onduty period:
  - These firefighters suffered heart attacks within 24 hours of onduty stressful or strenuous activity.
  - The timing of illness onset can range from when the firefighter is walking from the station to a vehicle after the conclusion of the incident, to the next day.
  - All of the firefighters included in this report under the Hometown Heroes Act criteria are in this group.
- Nine firefighters experienced heart attacks while on duty but not assigned to an incident, or participating in training:
  - Three firefighters suffered heart attacks while on duty in the fire station. Two of these firefighters were officers, and their illness struck as they worked in their offices.
  - Two New York firefighters became ill and died of heart attacks after marching in a fire department parade.
  - An Ohio firefighter suffered a heart attack at his home during a lunch break.
  - A North Carolina firefighter suffered a heart attack as he and his crew stood by for a community fireworks event.
  - A New Hampshire firefighter became ill while transporting a fire truck to another town for maintenance.
  - A North Carolina wildland firefighter suffered a heart attack while purchasing parts for a bulldozer.
- Eight firefighters died of heart attacks that struck while they were working on the scene of a fire incident:
  - Seven of the heart attacks struck as firefighters worked on the scene of structure fires.
  - One heart attack struck as the firefighter worked at the scene of a wildland fire.
- Seven firefighters suffered heart attacks while responding to an incident:
  - Three firefighters exhibited signs of heart attacks as they arrived at the fire station for a response. Two were transported to the hospital from the fire station and later died; one firefighter went home and became ill.
  - Two firefighters suffered heart attacks while driving vehicles. Their illnesses resulted in crashes. No other firefighters were hurt in the crashes.
  - A New York fire police officer experienced a heart attack during his response to the scene of an incident.
  - A Mississippi firefighter experienced a heart attack during a response and fell from the bed of a pickup truck.
- Five firefighters were involved in training activities when they had heart attacks:
  - Four firefighters experienced heart attacks during or shortly after physical fitness training activities.
  - A Virginia firefighter died of a cardiac-related illness during an extrication demonstration.
- Three firefighters became ill and died of heart attacks that struck while they were assigned to nonfire emergencies. All three occurred at motor vehicle crash scenes.

Internal Trauma

In 2006, 24 firefighters died due to internal physical trauma. This grouping includes most firefighters killed in vehicle crashes as well as those who receive physical injuries resulting from events such as a building collapse.

Table 12 – Internal Trauma Firefighter Deaths

Year	Number of Firefighter Deaths
2006	24
2005	32
2004	31
2003	41
2002	34
2001	28*
2000	36
1999	25
1998	27
1997	32

*Does not include the firefighter deaths of September 11, 2001, in New York City.

- Aircraft crashes accounted for seven traumatic firefighter deaths. The deaths resulted from three incidents:
  - Four wildland firefighters were killed in a helicopter crash in Idaho. The helicopter was engaged in a crew rotation mission when it crashed for reasons unknown.
  - Two California wildland firefighters were killed in the crash of a fixed-wing aircraft.
  - A California firefighter was killed in the crash of a helicopter during water drop operations at a wildland fire. The aircraft experienced a tail rotor failure and subsequent crash. The second occupant of the helicopter drowned.
- Motor vehicle-related incidents accounted for the deaths of 14 firefighters due to traumatic injuries:
  - Five firefighters were killed in crashes involving their personal vehicles. Three occurred during incident responses and two occurred during administrative duties.
  - Four firefighters were killed in crashes involving water tenders
  - Two firefighters were struck by fire department vehicles. One firefighter was run over by his own water tender as he worked at a prescribed burn; the other firefighter was run over by his apparatus when it was struck from behind by another vehicle.

- Two firefighters were killed when they were struck by vehicles. A Michigan firefighter was struck while working on an icy highway and a Pennsylvania firefighter was struck by a vehicle in a bank parking lot.
  - A New Mexico firefighter was killed during an emergency response in a county-owned pickup truck.
- A North Carolina firefighter was killed when a tornado struck his home as he departed for an emergency response.
- A South Dakota firefighter was killed when he was struck by hardware attached to a tow rope. When the rope attachment failed during an attempt to pull an apparatus from a field, the firefighter was struck by the recoiling rope and attachment.
- A Tennessee firefighter was struck by a falling tree at a prescribed burn.

Photo courtesy of Kimm Anderson and the St. Cloud Times.

Firefighter Kyle Weisbrich died in a motorcycle crash while responding to a motor vehicle crash.

Asphyxiation

Asphyxiation was the third leading medical reason for firefighter deaths in 2006. Twelve firefighters died due to asphyxiation, up from eight in 2005.

- Nine firefighters died as a result of inhaling the toxic products of combustion. All occurred during residential structural fires.
- Two firefighters died of drowning-related asphyxiation. One firefighter was the pilot of a helicopter that crashed into a body of water; the other was a firefighter who nearly drowned during surf rescue personal watercraft training and later died of complications from the near-drowning.
- A New York City firefighter died of asphyxiation when he was crushed by debris in a building collapse during a commercial structure fire.

Table 13 – Firefighter Deaths due to Asphyxiation

Year	Number of Firefighter Deaths
2006	12
2005	8
2004	5
2003	6
2002	15
2001	18
2000	13
1999	16
1998	15
1997	15

Burns

Eight firefighters died as a result of burns in 2006, up from three such deaths in both 2005 and 2004:

- Five California firefighters died of burns incurred when their position was overrun by rapidly advancing fire.
- Two firefighters died of burns received in wildland fires; one incident occurred in Oklahoma and the other in Utah.
- A Georgia firefighter was fatally burned in a residential structure fire.

Crushed

Five firefighters were crushed and killed in 2006. Four of these deaths occurred at the scene of structural fires, and one occurred at the fire station:

- Two Alabama firefighters were crushed in a wall collapse at a commercial structure fire.
- A New York fire officer was crushed in a building collapse. The officer survived the incident but died later.
- A Texas firefighter was killed when he was crushed in the collapse of a commercial structure under fire conditions.
- A New York firefighter was crushed between the roof of a heavy rescue fire apparatus and the fire station door frame as the apparatus was pulled out of the station. The firefighter was on the roof of the apparatus checking equipment and the driver of the apparatus was unaware of his location.

Cerebrovascular Accident

Four firefighters died in 2006 as a result of strokes (CVAs), down from six in 2005:

- A Texas career industrial firefighter suffered a CVA during training activities.
- A Texas firefighter suffered a CVA as he drove home from the fire station after going off duty.
- An Illinois firefighter experienced a CVA after she and her crew returned from an automatic fire alarm incident. She was transported to the hospital but her condition worsened and she died.
- A West Virginia firefighter suffered a CVA during a break from duties at the fire station.

Electrocution

A New York firefighter was electrocuted as he retrieved a salvage tarp that had been left on a commercial building that had experienced an earlier fire. The firefighter came into contact with an electrified sign that was not properly grounded.

Other

Two firefighters died in 2006 in situations where the nature of their deaths does not fall into any of the categories described above:

- A Mississippi firefighter died from a blood infection after cutting his finger on the scene of a motor vehicle crash. He died 7 days after the incident.
- A Pennsylvania firefighter suffered an aortic aneurysm while driving to the fire station to perform his assigned maintenance duties.

FIREFIGHTER AGES

Figure 13 shows the percentage distribution of firefighter deaths by age and nature of the fatal injury. Table 14 provides counts of firefighter fatalities by age and the nature of the fatal injury.

Figure 13. Fatalities by Age and Nature (2006)

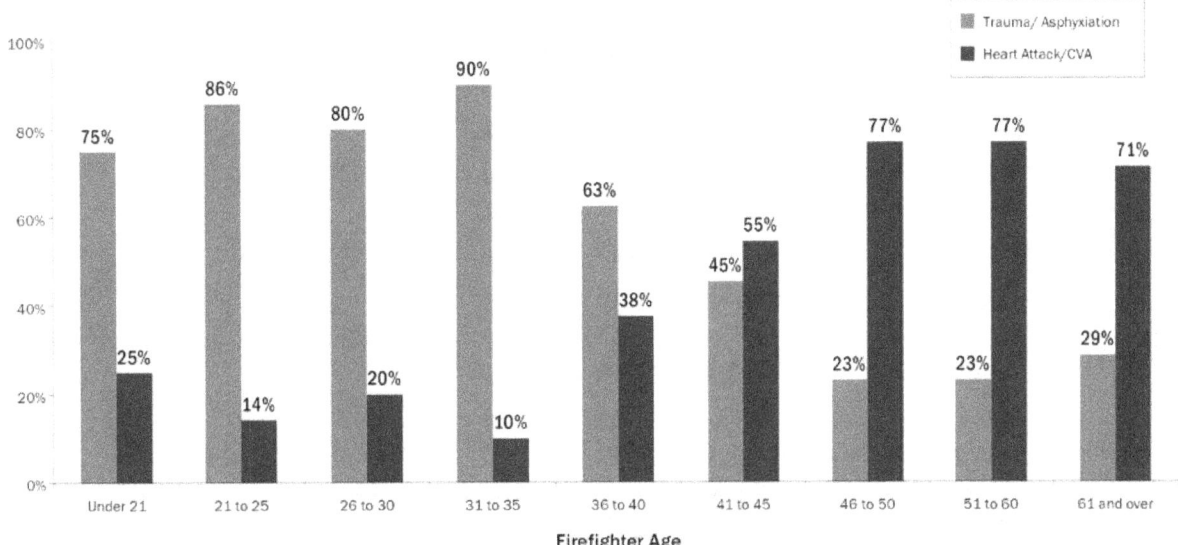

As in most prior years, younger firefighters were more likely to have died in 2006 as a result of traumatic injuries such as injuries from an apparatus accident or becoming caught or trapped during fire fighting operations. Although stress-related deaths are present in every age range in the 2006 data, stress plays an increasing role in firefighter deaths as age increases.

Table 14 – Firefighter Ages and Nature of Fatal Injury

Age Range	Non-Trauma Total	Trauma Total
under 21	1	3
21 to 25	1	6
26 to 30	1	4
31 to 35	1	9
36 to 40	6	10
41 to 45	6	5
46 to 50	10	3
51 to 60	20	6
61 & over	10	4

The youngest firefighter killed on-duty in 2006 was Firefighter Alethea Faye Nixon of Alabama. She was a passenger in a water tender crash at the age of 17. The oldest firefighter killed on-duty in 2006 was Firefighter John Stura of Pennsylvania. He died when he was struck by a motor vehicle at age 78.

FIXED PROPERTY USE FOR STRUCTURAL FIREFIGHTING DEATHS

There were 20 firefighter fatalities in 2006 where firefighters became ill while on the scene or engaged in structural firefighting and the fixed property use is known. Table 15 shows the distribution of these deaths by fixed property use. As in most years, residential occupancies accounted for the highest number of these fireground fatalities, with 15 deaths.

Table 16 shows the number of firefighter deaths in residential occupancies for the past 8 years. Residential occupancies usually account for 70 to 80 percent of all structure fires and a similar percentage of the civilian fire deaths each year*. Historically, the frequency of firefighter deaths in relation to the number of fires is much higher for nonresidential structures.

Table 15 – Structural Firefighting Deaths by Fixed Property Use (2006)

Fixed Property Use	Number	Percent
Residential	15	75%
Commercial	5	25%

Table 16 – Firefighter Deaths in Residential Occupancies

Year	Number of Firefighter Deaths
2006	15
2005	18
2004	15
2003	10
2002	21
2001	17
2000	21
1999	23
1998	17
1997	16

* Complete 2006 NFIRS fire incidence data were not available at the time of this report, but residential fires typically account for between 70 and 80 percent of all civilian fatalities each year, according to the NFPA.

TYPE OF ACTIVITY

In 2006, there were 36 firefighter deaths on the fireground. Table 17 and Figure 14 show the types of fireground activities firefighters were engaged in at the time they sustained their fatal injuries or illnesses. This total includes all firefighting duties such as wildland firefighting and structural firefighting.

Table 17. Type of Activity (2006)

Nature	Number
Fire Attack	24
Search and Rescue	4
Standby and Support	4
Pump Operations/Water	2
Incident Command	1
Unknown	1
Total	**36**

Figure 14. Fatalities by Type of Activity (2006)

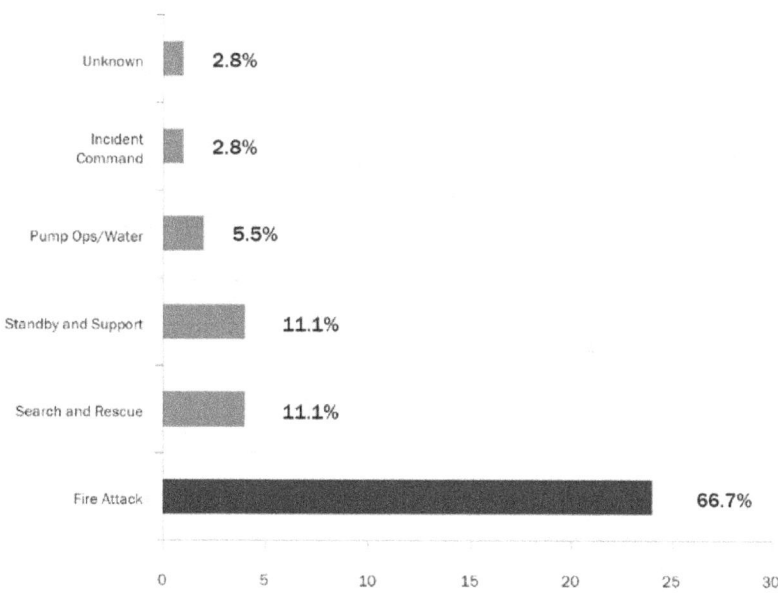

Note: Onscene fire only, 36 of 106 fatalities.

Fire Attack

In 2006, 24 firefighters were killed as they engaged in direct fire attack, such as advancing or operating a hoseline at a fire scene (see Table 18):

- Five California firefighters were killed as they defended a residence when fire progressed rapidly and overcame their position.
- Two California firefighters were killed when the tail rotor on their helicopter failed and the aircraft crashed into the body of water where they were drawing water at the time of the incident.
- Two Alabama firefighters were crushed when the wall of a commercial structure failed and collapsed on them.
- Two New York City firefighters were killed when the floor collapsed in a commercial building while the firefighters were inside fighting a fire.
- Four firefighters suffered heart attacks as they fought structural fires. The deaths occurred in Georgia, Illinois, North Carolina, and Pennsylvania.
- Four firefighters were killed when they became disoriented in residential structure fires. The incidents occurred in Colorado, Georgia, Indiana, and Mississippi.
- Two firefighters died during fire attack duties at wildland incidents. One firefighter was overrun by fire progress and the other was killed in a water tender rollover at a wildland incident.
- A Maryland firefighter was trapped in a residential structure fire by fire progress and a jammed door.
- A Texas firefighter was killed when a wall collapsed at a commercial building fire.
- An Indiana firefighter was killed when he fell through the fire-weakened floor of a residence and became trapped in the basement.

Table 18. Firefighter Deaths While Engaged in Fire Attack

Year	Number of Firefighter Deaths
2006	24
2005	11
2004	16
2003	11
2002	13
2001	13
2000	13
1999	16
1998	18
1997	21

With 24 deaths during fire attack, 2006 had the highest
number of such deaths in a decade.

Search and Rescue

Four firefighters were killed in 2006 as they engaged in search-and-rescue activities:

- Three of the deaths occurred in New Jersey. All three New Jersey fatal incidents took place in residential structures.
- A Wisconsin firefighter was killed when the floor in a residential structure failed during search-and-rescue activities. The firefighter was unable to escape the basement.

Standby and Support

Four firefighters died in 2006 while performing standby or support duties at the fire scene:

- Two California firefighters were killed in the crash of a fixed-wing aircraft. The firefighters were assigned to perform aerial observation duties.
- A Utah command officer was killed when he was overrun by fire progress at a wildland fire. The firefighter was assigned as a member of the incident management team.
- A South Dakota firefighter performing support duties at a wildland fire was killed when he was struck by an object as firefighters attempted to pull a fire apparatus from a field.

Pump Operations – Water Supply

Two firefighters died in 2006 while engaged in water supply duties:

- A Pennsylvania firefighter suffered a heart attack as he staffed a pump panel at a mutual-aid structure fire.
- A South Carolina firefighter performing water supply duties collapsed and died at the scene of a structure fire.

Incident Command

A New York command officer collapsed and died on the scene of a structure fire in a neighboring town.

Other Activity

One firefighter died of a heart attack on the scene of a wildland fire incident. His activities on scene are undetermined.

TIME OF INJURY

The distribution of all 2006 firefighter deaths according to the time of day when the fatal injury occurred is illustrated in Figure 15. The time of fatal injury for four firefighters was either unknown or unreported.

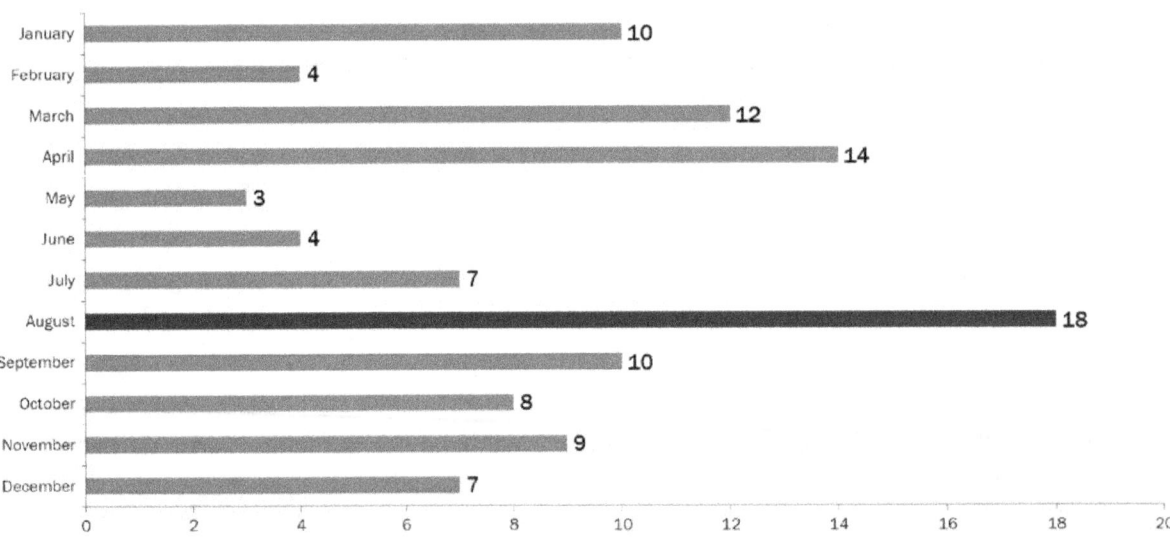

Figure 15. Fatalities by Time of Fatal Injury (2006)

Time	Fatalities
0100 - 0259	7
0300 - 0459	4
0500 - 0659	7
0700 - 0859	10
0900 - 1059	9
1100 - 1259	9
1300 - 1459	15
1500 - 1659	6
1700 - 1859	10
1900 - 2059	10
2100 - 2259	11
2300 - 0059	4

MONTH OF THE YEAR

Figure 16 illustrates the 2006 firefighter fatalities by month of the year.

Figure 16. Deaths by Month of the Year (2006)

Month	Deaths
January	10
February	4
March	12
April	14
May	3
June	4
July	7
August	18
September	10
October	8
November	9
December	7

STATE AND REGION

The distribution of firefighter deaths in 2006 by State is shown in Table 19*. Firefighters based in 36 States died in 2006.

The highest number of firefighter deaths based on the location of the fire service organization in 2006 occurred in New York with 14 deaths. New York also had the highest number of onduty firefighter deaths in 2005, with 18 deaths.

Table 19. Firefighter Fatalities by State by Location of Fire Service (2006)

Number	State	Percent of Total
3	Alabama	2.83%
11	California	10.3%
1	Colorado	0.94%
1	Connecticut	0.94%
4	Georgia	3.77%
3	Idaho	2.83%
4	Illinois	3.77%
3	Indiana	2.83%
2	Kansas	1.88%
2	Kentucky	1.88%
2	Maryland	1.88%
1	Michigan	0.94%
1	Minnesota	0.94%
1	Missouri	0.94%
4	Mississippi	3.77%
11	North Carolina	10.3%
1	Nebraska	0.94%
2	New Hampshire	1.88%

continued on next page

* This list attributes the deaths according to the State in which the fire department or unit is based, as opposed to the State in which the death occurred. They are listed by those States for statistical purposes and for the National Fallen Firefighters Memorial at the National Emergency Training Center.

Number	State	Percent of Total
5	New Jersey	4.71%
1	New Mexico	0.94%
1	Nevada	0.94%
14	New York	13.2%
3	Ohio	2.83%
2	Oklahoma	1.88%
1	Oregon	0.94%
6	Pennsylvania	5.66%
1	Rhode Island	0.94%
2	South Carolina	1.88%
1	South Dakota	0.94%
3	Tennessee	2.83%
4	Texas	3.77%
1	Utah	0.94%
1	Virginia	0.94%
1	Washington	0.94%
1	Wisconsin	0.94%
1	West Virginia	0.94%

Figure 17. Firefighter Fatalities by Region (2006)

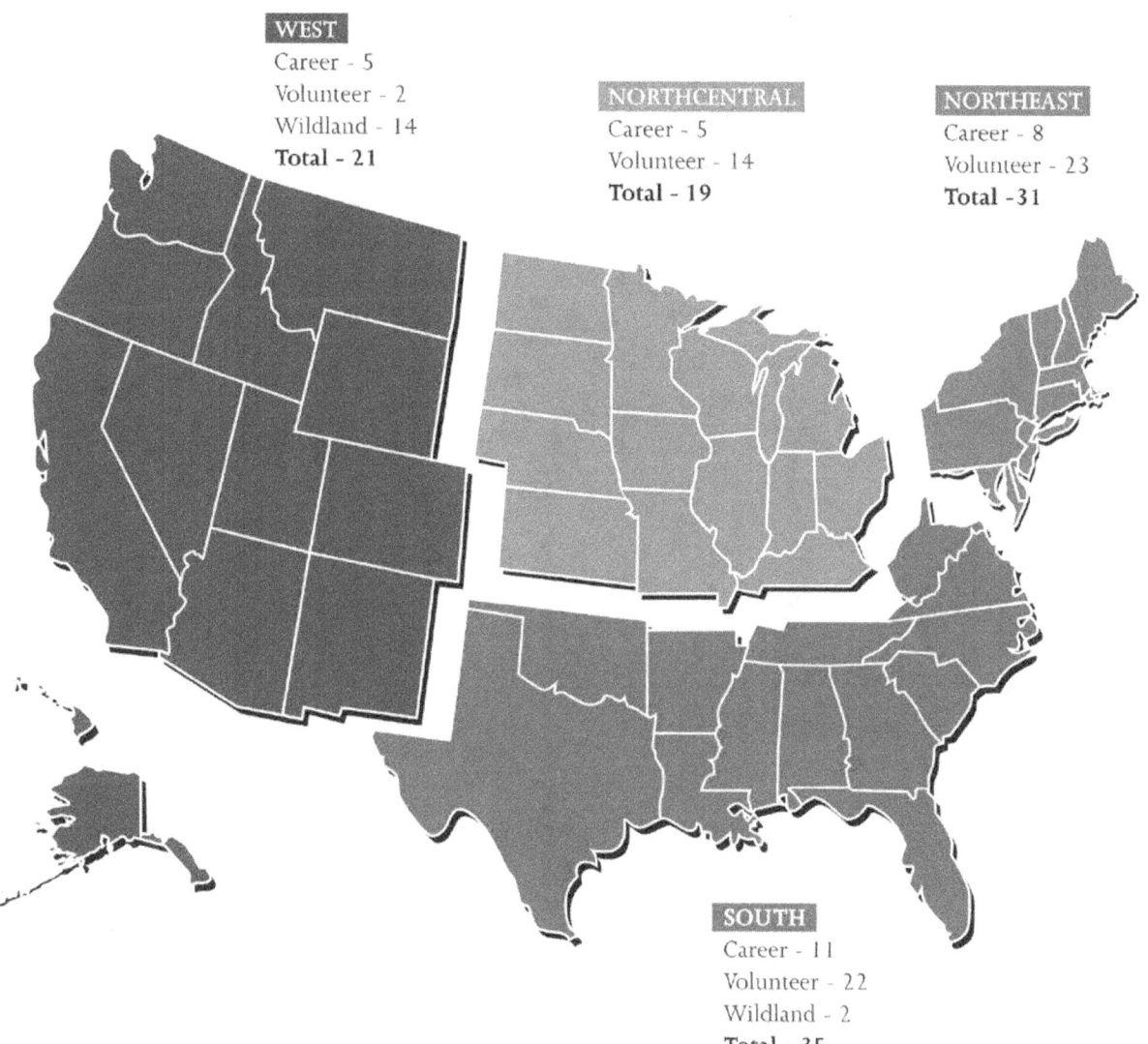

WEST
Career - 5
Volunteer - 2
Wildland - 14
Total - 21

NORTHCENTRAL
Career - 5
Volunteer - 14
Total - 19

NORTHEAST
Career - 8
Volunteer - 23
Total -31

SOUTH
Career - 11
Volunteer - 22
Wildland - 2
Total - 35

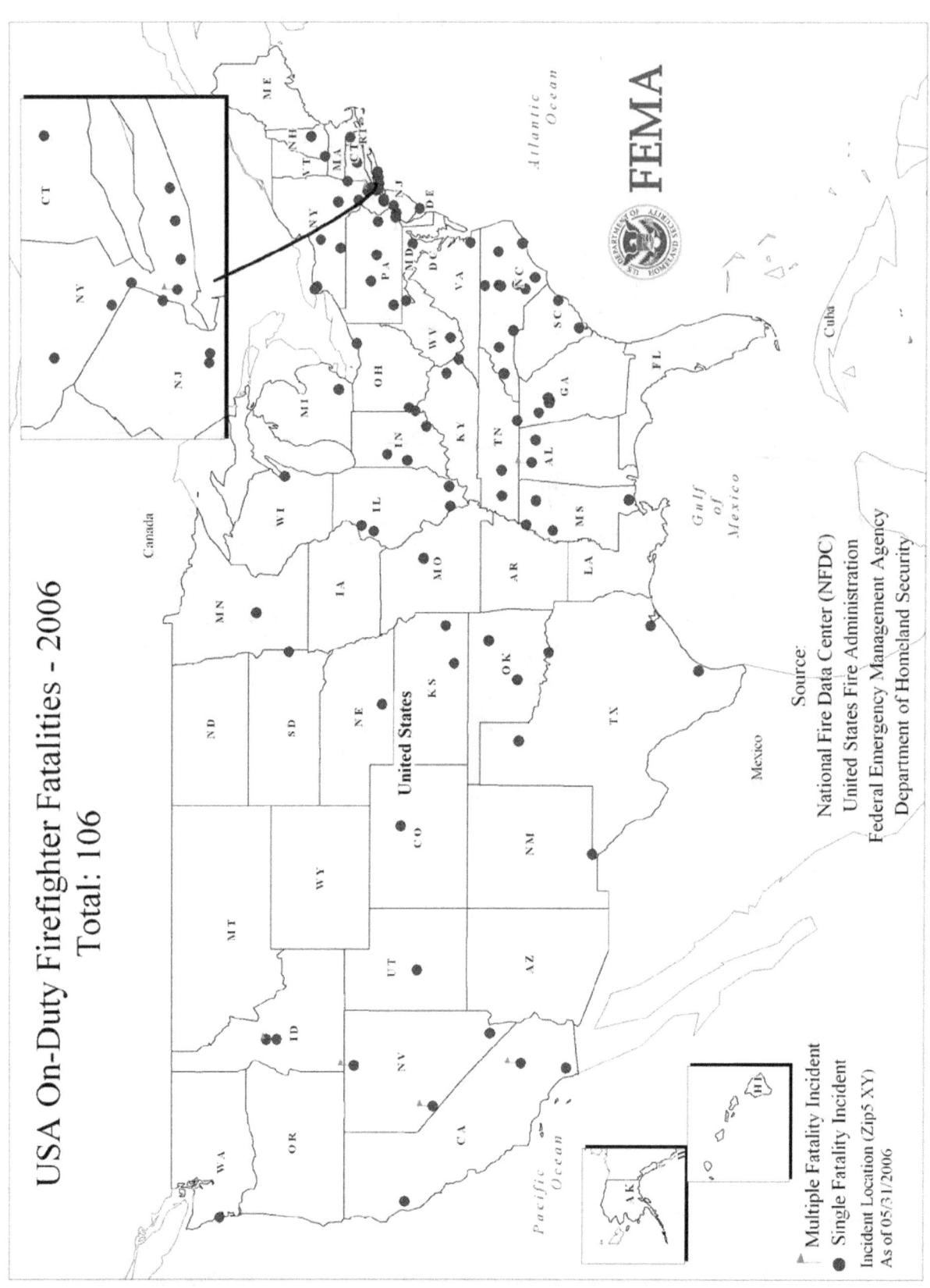

USA On-Duty Firefighter Fatalities - 2006
Total: 106

Multiple Fatality Incident
Single Fatality Incident
Incident Location (Zip5 XY)
As of 05/31/2006

Source:
National Fire Data Center (NFDC)
United States Fire Administration
Federal Emergency Management Agency
Department of Homeland Security

Analysis of Urban/Rural/Suburban Patterns in Firefighter Fatalities

The United States Census Bureau defines "urban" as a place having a population of at least 2,500 or lying within a designated urban area. Rural is defined as any community that is not urban. Suburban is not a census term but may be taken to refer to any place, urban or rural, that lies within a metropolitan area defined by the Census Bureau, but not within one of the central cities of that metropolitan area.

Fire department areas of responsibility do not always conform to the boundaries used by the Census Bureau. For example, fire departments organized by counties or special fire protection districts may have both urban and rural coverage areas. In such cases, where it may not be possible to characterize the entire coverage area of the fire department as rural or urban, firefighter deaths were listed as urban or rural based on the particular community or location in which the fatality occurred.

The following patterns were found for 2006 firefighter fatalities. These statistics are based on answers from the fire departments and, when no data from the departments were available, the data are based upon population and area served as reported by the fire departments.

Table 20. Firefighter Deaths by Coverage Area Type

	Urban/Suburban	Rural	Federal or State Parks/Wildland	Total
Firefighter Deaths	44	46	16	106

The last firefighter fatality in 2006 was Firefighter Phillip Wayne Townsend of the Denison Fire Department in Texas. Firefighter Townsend was 31 years old and he was killed in a structural collapse at a commercial building fire

IN CONCLUSION

In 2006, the fire service family in the United States suffered another year of unacceptable death and injury.

The year brought some hopeful signs, a relatively substantial single-year dip in the total number of firefighter fatalities that with continued daily leadership, teamwork, and focused effort by each and every one of the more than one million firefighters in the USA, can become a multiyear trend. As with any year, the sacrifice and losses of 2006 must not be forgotten. The lessons that we learn from these tragic incidents serve the fire service best if they are used as the basis of preventing future deaths.

In 2006, many fire service organizations made major efforts to spread the word about firefighter safety and to change the basic culture of firefighting to one that considers risk more actively. These efforts seem to be bearing fruit and they continue.

There are many facets to the firefighter fatality problem. Some fatalities could be avoided by actions as simple as slowing down the speed of a response or the wearing of a seatbelt. Other problems, such as building construction and firefighter health issues, require a more comprehensive and longer-term solution.

No mission is more important to the fire service than the elimination of needless firefighter deaths. This document seeks to support this effort in some small way.

In memory of the firefighters who answered their last call in 2006
To their friends and families
To their service and sacrifice

Special Topic

Engineered Wood Products

Sunday, June 25, 2006—1432hrs—Lafayette, Indiana

Deputy Chief Steven A. Smith and members of the Wea Township Volunteer Fire Department were dispatched to a report of a structure fire after lightning struck a house and ignited a basement fire.

Deputy Chief Smith was the first firefighter on the incident scene and found a working fire. Wearing full structural firefighting protective clothing and a self-contained breathing apparatus (SCBA), Deputy Chief Smith and another firefighter entered the structure and found the floor to be spongy. Deputy Chief Smith ordered the other firefighter to apply water to the basement through a side window as he advanced an attack line through the front door of the residence. Immediately upon re-entering the structure, Deputy Chief Smith fell into the basement through the fire-weakened floor.

Other firefighters arriving on the scene found the attack line leading into the basement. Deputy Chief Smith called to firefighters from the basement and told them he was trapped and needed rescue. Firefighters attempted to enter the basement through the hole using a ground ladder, but were unable to make access.

The homeowner directed firefighters to the basement stairway and two firefighters entered the basement to search for Deputy Chief Smith. Firefighters followed the sound of Deputy Chief Smith's PASS device and located him; he was unconscious and his facepiece was not in place. Firefighters began dragging Deputy Chief Smith toward the stairs, but he became entangled in debris and firefighters attempted to remove his SCBA to provide air from their SCBA units.

Deputy Chief Smith was removed from the building and transported to the hospital by ambulance; he was pronounced dead after his arrival. The cause of death was listed as smoke inhalation. At autopsy, Deputy Chief Smith's carboxyhemoglobin level was 57 percent.

The primary structural component of the flooring system that collapsed from exposure to the fire was an engineered wood product consisting of solid wood top and bottom plates and a chipboard web. This section will provide a brief overview of engineered wood products with recommendations for safe firefighting operations and resources for additional information.

Overview

While not a new phenomenon, the prevalence of lightweight structural components in wall, floor, and roof assemblies is an area of continued fire service concern. For many applications, lightweight wood structural components including structural insulated panels (SIPs), glue laminated beams ("glulams"), I-joists, structural composite lumber (SCL), and trusses have replaced dimensional lumber in building construction. Engineered wood products (EWPs) such as wooden I-joists, SCL, and trusses are commonly used in floor and roof assemblies across the United States. According to the American Wood Council (AWC), part of the American Forest & Paper Association (AF&PA):

Photo courtesy of the Wea Township Volunteer Fire Department.

Deputy Chief Steven Smith died when he fell into the basement of a residential structure fire.

> Engineered wood products can be defined as products consisting of a combination of smaller components to make a structural product, designed using engineering methods. They are an alternative to traditional sawn lumber. To use as much fiber as possible from available logs, the logs are "stranded" (sliced into 1-inch to 12-inch strands), peeled (sliced off of a log like paper towels from a roll), or sawn. Then, the resulting pieces are combined with adhesives into a new engineered wood product. (AF&PA, *www.woodaware.info/guideewp.html*, 2007)

Used in most new residential construction, during the renovation of many older buildings, and in some commercial applications, floor and roof assemblies constructed using engineered wood products are proven under normal conditions; however, their performance during a fire depends on a number of factors where manufacturers and firefighters have little control.

In 2006, two firefighters were killed in structure fire incidents where the failure of fire-damaged engineered wood products was cited as a contributing factor.

- An Indiana volunteer firefighter died after falling into the basement of a residential structure when fire-weakened flooring gave way.
- A Wisconsin firefighter fell through the fire-weakened floor of a residential structure and was trapped in the fully-involved basement.

Given the uncertain performance of EWPs under fire conditions, firefighters must quickly recognize the presence of these building components and make appropriate adjustments to firefighting strategy and tactics.

Before the Fire

Firefighters can take a number of proactive steps that are important for operating safely around engineered wood products. Maintaining familiarity with the response area is critical to identify structures built or renovated using EWPs; these structures can be noted on response maps, preincident plans, and during company/department briefings or training sessions. Firefighters can work with building safety and code enforcement personnel to help ensure that EWPs are used and installed according to manufacturers' recommendations. In jurisdictions that have labeling requirements for buildings with truss systems, consideration also can be given to labeling buildings with EWPs such as floors constructed using wooden I-joists. The installation and maintenance of a residential (or commercial) fire sprinkler system is perhaps the best way to protect the structural integrity of EWPs and ensure the safety of building occupants and firefighters.

Strategic Considerations

Operating inside a burning structure is always dangerous, regardless of any specific building construction practices or materials. Firefighters and incident managers must weigh the potential risks and benefits of taking an offensive strategy quickly and carefully before committing to interior operations. The IAFC has developed "10 Rules of Engagement for Structural Firefighting" that apply to all fires:

Acceptability of Risk

1. No building or property is worth the life of a firefighter.
2. All interior fire fighting involves an inherent risk.
3. Some risk is acceptable, in a measured and controlled manner.
4. No level of risk is acceptable where there is no potential to save lives or savable property.
5. Firefighters shall not be committed to interior offensive fire fighting operations in abandoned or derelict buildings.

Risk Assessment

1. All feasible measures shall be taken to limit or avoid risks through risk assessment by a qualified officer.
2. It is the responsibility of the Incident Commander to evaluate the level of risk in every situation.
3. Risk assessment is a continuous process for the entire duration of each incident.
4. If conditions change, and risk increases, change strategy and tactics.
5. No building or property is worth the life of a firefighter.

   (IAFC, *www.iafc.org/associations/4685/files/rules.pdf*, 2001)

When direct fire impingement on EWPs used in wall, floor, or roof construction is known or suspected, a defensive strategy is likely the safest course of action, depending on the location, intensity, and duration of the fire. Exercise special caution in cases where fire is located beneath unprotected EWP assemblies, such as basement or attic fires; under these circumstances EWPs may fail unexpectedly and without any warning signs. Assuming or maintaining an offensive strategy when EWPs are involved in fire is an extremely high-risk situation that must be factored into the incident action plan (IAP).

Tactical Considerations

As with any structure fire, sizeup for buildings constructed using EWPs must be a continual process that evaluates a wide range of factors. Completing a 360-degree "lap" of the building should be a priority to identify the presence of fire in the basement, crawlspace, attic, or trussloft. Special emphasis should be placed on rapidly identifying the existence of EWPs in floor/roof assemblies and whether or not the EWPs have been exposed to heat or flame; this information must be communicated to the Incident Commander. Checking overhead spaces and sounding floors immediately upon entering a fire building is a sound tactical practice; firefighters should be aware, however, that floor/roof assemblies constructed with EWPs can fail without exhibiting any warning signs such as charring or feeling weak and "spongy." In addition to physical inspection, the routine use of a thermal imaging camera (TIC) to sweep the floor and overhead for signs of heat and fire is highly recommended; the presence of certain floor coverings such as tile or carpeting can mask returns from the TIC.

Resources

Under a cooperative agreement with the USFA, the AF&PA developed a series of eight Awareness Guides covering different wood and engineered wood products of great importance to firefighters:

- Engineered Wood Products Primer
- I-Joists
- Trusses
- Glulam/Structural Composite Lumber
- Panels
- Lumber
- Adhesives
- Fire Performance of Wood Products

The eight Awareness Guides, along with other educational materials such as the interactive FireFrame building construction tool, are available directly from the AF&PA using the following World Wide Web address: *www.woodaware.info.*

The USFA Web site on building performance awareness of lightweight construction during fires may be accessed using: *www.usfa.dhs.gov/fireservice/research/safety/construction.shtm*

The AF&PA and the Illinois Fire Service Institute (IFSI) have produced a firefighter awareness training program on wooden structural components that is available from the IFSI through its "Virtual Campus" Web site at: *www.fsi.uiuc.edu.*

Photo by Mark A. Whitney

The National Fallen Firefighters Foundation's nationwide "Whistlestop Tour" promoting firefighter safety issues, with a key message of "Everyone Goes Home," stopped at the National Emergency Training Center in Emmitsburg, Maryland in April, 2007.

www.EveryoneGoesHome.com

Appendix A
Summary of 2006 Incidents

January 3, 2006 – 0839hrs
Richard O. Longoria, Firefighter II/Paramedic
Age 54, Career
Corpus Christi Fire Department, Texas

Firefighter Longoria completed a 24-hour shift. During the shift, he responded to a lift-assist incident. Prior to leaving the fire station, Firefighter Longoria complained of not feeling well to other firefighters.

Firefighter Longoria left the fire station enroute to his home. When he was approximately 7.6 miles from the fire station, Firefighter Longoria suffered a CVA (stroke). He was found by a passerby lying in the street near his car. Firefighters responded to the incident and provided Firefighter Longoria with emergency medical care. Despite their efforts, Firefighter Longoria was pronounced dead at the scene by a representative of the county medical examiner's office.

January 7, 2006 – 0708hrs
Amy Lynn Schnearle-Pennywitt, Firefighter
Age 34, Career
Ann Arbor Fire Department, Michigan

Firefighter Schnearle-Pennywitt and her engine company were on the scene of a multiple-vehicle weather-related crash. A light drizzle was falling and black ice conditions were reported on the scene. The incident scene was located in an area that was not visible to drivers until they were almost upon it. Firefighter Schnearle-Pennywitt was wearing structural firefighting protective clothing.

As she and her officer walked the scene to assess the severity of injuries to drivers of the vehicles involved in the crash, a Ford F150 pickup truck that was not involved in the initial incident lost control, struck the median wall, struck a vehicle, and then struck Firefighter Schnearle-Pennywitt from behind. The company officer was also struck but was able to come to Firefighter Schnearle-Pennywitt's side to provide assistance. As additional firefighters arrived on the scene, they discovered Firefighter Schnearle-Pennywitt unconscious, on her side in a ditch. Firefighters provided emergency medical care and prepared her for transport.

Firefighter Schnearle-Pennywitt was transported by ambulance to a regional trauma facility. Due to the nature and severity of her injuries, Firefighter Schnearle-Pennywitt was in a coma. She was removed from life support and died on January 13, 2006. The cause of death was listed as brain trauma.

January 11, 2006 – 1153hrs
David Robert Packard, Lieutenant
Age 41, Volunteer
Swanzey Fire Department, New Hampshire

Lieutenant Packard and the members of his fire department responded to an emergency medical incident on January 10, 2006. The incident involved an 85-year-old man in cardiac arrest. Lieutenant Packard completed his duties at 1730hrs.

The next morning, Lieutenant Packard ate breakfast with his family, delivered a bid to a customer, and arrived at a plumbing job site. At 1153hrs, Lieutenant Packard felt ill, called his wife to tell her that he was going to the hospital, and was driven to the hospital by a friend. During travel to the hospital, Lieutenant Packard became unconscious. Upon his arrival at the hospital, he was found to be in cardiac arrest.

Hospital emergency room staff attempted to resuscitate Lieutenant Packard for 45 minutes without success. He was pronounced dead at 1300hrs. The cause of death was listed as coronary atherosclerosis.

January 14, 2006 – 1200hrs
Jason Allen Johnson, Firefighter
Age 28, Volunteer
Butler County Fire District #3, Kansas

Firefighter Johnson responded to two emergency incidents on the morning of January 14, 2006. His response to the first emergency ended at the fire station when the incident response was cancelled. Firefighter Johnson's second emergency response was to a wildland fire where he performed very strenuous tasks for approximately 1 hour.

After returning home and cleaning up from the wildland fire, Firefighter Johnson left home with members of his family to complete errands. Less than an hour after returning from his second emergency incident, Firefighter Johnson collapsed in his personal vehicle and became unresponsive.

Firefighter Johnson was removed from his vehicle, CPR was initiated, and EMS was called. Upon the arrival of EMS responders, an Automatic External Defibrillator (AED) was applied but advised that no shock was indicated. Paramedics arrived and provided ALS (Advanced Life Support) level assistance. Firefighter Johnson was transported to the hospital. Despite all of these efforts, Firefighter Johnson was pronounced dead at 1626hrs.

Firefighter Johnson's death was likely caused by a cardiac condition. For additional information regarding this incident, please refer to NIOSH Fire Fighter Fatality Investigation and Prevention Program report F2006-13 (*www.cdc.gov/niosh/fire/reports/face200613.html*).

January 20, 2006 – 2336hrs
Roger W. Armstrong, Firefighter
Age 42, Volunteer
Atlanta Fire Department, Illinois

Firefighter Armstrong and the members of his fire department were engaged in fighting a mutual-aid house fire. When his SCBA air supply was depleted, Firefighter Armstrong removed the SCBA and sat on the running board of one of the fire apparatus on scene. Firefighter Armstrong complained of not feeling well and was directed to EMS responders on the scene of the incident.

Firefighter Armstrong was treated at the scene, his condition worsened, and he was transported to the hospital. Firefighter Armstrong was treated at the hospital but was not revived. He was pronounced dead at 0135hrs on January 21, 2006. The cause of death was listed as a coronary thrombosis (blocked artery).

❧

January 22, 2006 – 1405hrs
John Robert "Bobby" Westervelt, Firefighter
Age 57, Volunteer
Spring Valley Fire Department, New York

Firefighter Westervelt and the members of his fire department responded to their fire station for an automatic fire alarm in a senior citizens housing facility. Prior to leaving the fire station, the incident was found to be false and the response was cancelled.

Firefighter Westervelt was discovered by firefighters outside of the fire station. He was unresponsive and emergency medical care was initiated. Efforts to revive Firefighter Westervelt at the fire station and after his arrival at the hospital were unsuccessful. His death was caused by a heart attack.

❧

January 23, 2006 – 1330hrs
Jack Eugene Arnold, Forest Fire Equipment Operator
Age 48, Wildland Full-Time
North Carolina Division of Forest Resources, North Carolina

Forest Fire Equipment Operator (FFEO) Arnold was on duty the morning of January 23, 2006. He spent the morning working on fire control bulldozers and preparing for an upcoming equipment inspection. Specifically, he removed the belly pans on a bulldozer. This work involved working on his back and maneuvering a 125-pound pan in the area beneath the bulldozer.

Immediately after lunch, FFEO Arnold and another firefighter drove to pick up parts at a local automotive parts distributor. FFEO Arnold waited in the vehicle while the other firefighter went into the store. When the firefighter came back to the vehicle, he found FFEO Arnold slumped over the wheel of the truck. FFEO Arnold was unresponsive and the EMS system was activated. Paramedics arrived and transported FFEO Arnold to the hospital where he was later pronounced dead.

The cause of death was listed as severe coronary artery disease.

❧

January 26, 2006 – 0918hrs
Harold Vernon Taylor, Assistant Chief
Age 65, Volunteer
Central Warren County Fire Protection District, IL

Assistant Chief Taylor and the members of his fire department were dispatched to a structure fire in a residence. When Assistant Chief Taylor arrived at the fire station, he told other firefighters that he had awakened with a severe headache and that he was not going to respond to the incident. The firefighters expressed their concern for Assistant Chief Taylor's well-being but were told by Assistant Chief Taylor to respond to the incident.

After the firefighters departed, Assistant Chief Taylor returned to his residence to change clothes;, he had vomited. When Assistant Chief Taylor arrived home, he collapsed. His wife activated the 9-1-1 system and an ambulance was dispatched to the residence. Assistant Chief Taylor was transported to the hospital but did not survive.

The cause of death was listed as a heart attack.

January 28, 2006 – 0430hrs
Tracy Champion, Firefighter
Age 49, Career
Philadelphia Fire Department, Pennsylvania

Firefighter Champion and the members of his engine company were first on the scene of a working fire in a rowhouse. Firefighter Champion and another firefighter, while wearing full structural firefighting protective clothing and SCBA, stretched a handline to the rear of the residence in order to fight the fire.

After approximately 2 hours on the scene, Firefighter Champion and other firefighters were wetting down debris on the first floor of the structure when Firefighter Champion suddenly collapsed. A medic unit that had recently departed the scene was recalled.

Firefighters removed Firefighter Champion from the structure and began emergency medical care. CPR was initiated and an AED was applied as paramedics arrived on the scene. A manual defibrillator was used to deliver three shocks in an attempt to correct Firefighter Champion's heart rhythm.

Firefighter Champion was transported to the hospital and care was continued in the emergency room. Firefighter Champion was pronounced dead at 0525hrs, 55 minutes after his collapse.

The cause of death was listed as atherosclerotic cardiovascular disease.

For additional information regarding this incident, please refer to NIOSH Fire Fighter Fatality Investigation and Prevention Program report F2006-09 (www.cdc.gov/niosh/fire/reports/face200609.html).

January 30, 2006 – 2220hrs
Gary Wayne Kistler, Sr., Firefighter
Age 65, Volunteer
Saucier Volunteer Fire Department, Mississippi

Firefighter Kistler responded to the scene of a motor vehicle crash on January 30, 2006. During the course of work on the scene of the incident, Firefighter Kistler cut his finger. Firefighter Kistler cleaned the wound and continued working on the incident.

The day after the incident, Firefighter Kistler complained of not feeling well. Two days after the incident, Firefighter Kistler went to the hospital complaining of pain is his shoulder blade area and a rash in the area of complaint. He was given muscle relaxants and sent home from the hospital.

Five days after the incident, Firefighter Kistler returned to the hospital. While in the emergency room, Firefighter Kistler suffered cardiac arrest. He was resuscitated and sent to the hospital's intensive care unit. On February 5, 2006, 6 days after he was injured, Firefighter Kistler died.

The cause of death was listed as septicemia, a blood infection.

February 9, 2006 – 0543hrs
Edward Joseph Marbet, Firefighter
Age 31, Volunteer
Burlington Township Fire Department, New Jersey

Firefighter Marbet was asleep in his residence when he was awakened by a smoke alarm. He rose from the bed to investigate the source of the alarm. Shortly thereafter, he called to his fiancée and requested that she bring him water. Firefighter Marbet called 9-1-1 to report the fire. He then escorted his fiancée out of the house and closed the door behind her. Firefighter Marbet also told his fiancée to make sure that the other occupant of the residence was removed.

After Firefighter Marbet's fiancée placed a pet into a car in the driveway, she heard Firefighter Marbet calling for help from behind a door. She was unable to open the door.

Fire investigators concluded that a fire occurred in a couch. Firefighter Marbet attempted to extinguish the fire in the couch and remove it from the residence. Firefighter Marbet was able to get the couch down a flight of stairs that led to an exit, but the couch became lodged in the stairway. The couch began to burn freely and cut off Firefighter Marbet's escape route. He was able to retreat to an upper floor of the house, but collapsed prior to reaching a window.

Firefighters arriving on the scene found a well-developed fire. After the fire was extinguished, the body of Firefighter Marbet was located. An autopsy listed the cause of death as smoke inhalation.

February 21, 2006 – 2150hrs
Lloyd B. McCulloch, Captain
Age 63, Volunteer
Moulton Fire Department, Alabama

Dustin Keith Jones, Firefighter
Age 23, Volunteer
Moulton Fire Department, Alabama

Captain McCulloch and Firefighter Jones responded with other members of their fire department to a structure fire in a single-story commercial building. The fire fight involved a number of fire departments and stretched over a period of hours.

Approximately 4 hours after the initial alarm, Captain McCulloch, Firefighter Jones, and another firefighter stretched a handline to the front of the store to extinguish hot spots. As the firefighters worked, an overhead awning collapsed inward toward the firefighters, and the front wall of the building collapsed outward. Captain McCulloch and Firefighter Jones were buried in debris.

Firefighters on the scene summoned help and attempted to remove debris and locate the trapped firefighters. Airbags were used to lift portions of the collapsed building and both trapped firefighters were located approximately 20 minutes after the collapse. Both firefighters were pronounced dead at the scene.

The cause of death for both firefighters was listed as multiple blunt force trauma.

For additional information regarding this incident, please refer to NIOSH Fire Fighter Fatality Investigation and Prevention Program report F2006-07 (www.cdc.gov/niosh/fire/reports/face200607.html).

February 22, 2006 – 1810hrs
Robert "Ockie" Wisting, Firefighter
Age 77, Volunteer
Middle Township Fire District #2, Rio Grande Volunteer Fire Company, New Jersey

Firefighter Wisting and the members of his fire department were at their fire station engaged in a weekly drill. A passerby stopped at the station and reported a smoke condition on a street near the fire station. Firefighters responded and found that the source of the smoke was stacks at a local generating plant. The incident was concluded and firefighters returned to the fire station by 2051hrs.

Firefighter Wisting complained to other firefighters that he did not feel well. He went home and slept until 1500hrs the next day. Firefighter Wisting took his dog outside to his yard. While in the yard, Firefighter Wisting collapsed of a sudden heart attack.

Firefighters and rescue squad personnel responded and provided treatment for Firefighter Wisting. He was transported to the hospital but was later pronounced dead. The cause of death was listed as hypertensive atherosclerotic cardiovascular disease.

March 1, 2006 – 1657hrs
John Destry Horton, Firefighter/Paramedic
Age 32, Volunteer
Acme Fire Department, Chickasha, Oklahoma

Firefighter/Paramedic Horton and the members of his fire department were dispatched to a wildland fire as mutual-aid to the Duncan Fire Department. Upon their arrival on the scene, Firefighter/Paramedic Horton and another Acme firefighter were assigned to attack a portion of the fire.

Firefighter/Paramedic Horton drove the brush apparatus from a burned portion of land to an unburned area to access another approach to the fire. Firefighter/Paramedic Horton was wearing structural firefighting protective trousers but no other protective clothing. The apparatus was being turned around when a sudden wind shift occurred. As he attempted to drive to an area of safety, the fire overtook Firefighter/Paramedic Horton's apparatus. Firefighter/Paramedic Horton left the cab of the apparatus to provide assistance to the other Acme firefighter. In the process, his helmet fell off and he received severe upper body burns, burns to his feet, and respiratory burns.

Firefighter/Paramedic Horton was transported by medical helicopter to a regional burn center in Oklahoma City. Firefighter/Paramedic Horton died as a result of his burns on March 24, 2006 at 2130hrs. The cause of death was listed as multisystem organ failure.

In addition to his membership in the Acme Fire Department, Firefighter/Paramedic Horton was a career firefighter with the Chickasha Fire Department.

ᔓ

March 3, 2006 – 2116hrs
Robert John Schnibbe, Jr., Battalion Chief
Age 57, Part-Time (Paid)
Hastings-on-Hudson Volunteer Fire Department, New York

Battalion Chief Schnibbe and the members of his fire department were dispatched to assist with a working residential structure fire in the Village of Irvington. In his role as mutual-aid coordinator, Battalion Chief Schnibbe responded directly to the scene while other members of his department stood by for coverage at the Irvington fire station.

Battalion Chief Schnibbe was leaving the scene to return to his car when he suddenly collapsed. Emergency medical care was provided and Battalion Chief Schnibbe was transported to the hospital. Battalion Chief Schnibbe was pronounced dead at the hospital. The cause of death was listed as occlusive coronary atherosclerosis.

ᔓ

March 5, 2006 – 1720hrs
Wayne Edward "Butch" Yarborough, Lieutenant
Age 59, Volunteer
Waynesville Fire Department, North Carolina

Lieutenant Yarborough and the members of his fire department responded to a small grass fire behind an apartment complex. The fire appeared to be a controlled burn that was left unattended. The fire was extinguished with a 1-1/2-inch handline. The total burned area was approximately 100 square feet.

continued on next page

Lieutenant Yarborough assisted with extinguishment of the fire and helped to repack the hoseline back onto the apparatus. After the apparatus returned to the fire station, Lieutenant Yarborough returned to his home.

Shortly after eating dinner, Lieutenant Yarborough collapsed. Family members called 9-1-1. Arriving responders provided CPR and Lieutenant Yarborough was transported to the hospital, where he was later pronounced dead.

The cause of death was listed as hypertensive atherosclerotic cardiovascular disease.

March 8, 2006 – 0045hrs
Michael Lynn Davenport, Inmate Firefighter
Age 39, Volunteer
Mississippi Department of Corrections - Parchman Volunteer Fire Department, Mississippi

Firefighter Davenport was a member of the volunteer fire department of the Mississippi State Penitentiary at Parchman. He and the members of his fire department were dispatched to a residential structure fire in the community.

Firefighters discovered a fire in the attic of a split-level house. Firefighter Davenport and other firefighters worked on the second floor of the structure and attempted to access the fire. Firefighters were driven from the house due to intense heat. Firefighter Davenport became disoriented in the structure and collapsed of smoke inhalation.

⌒

March 11, 2006 – Time Unknown
Jeffrey A. Bowman, Lieutenant
Age 42, Career
Chattanooga Fire Department, Tennessee

Lieutenant Bowman worked a 24-hour shift beginning at 0700hrs on March 9, 2006. During the shift, he responded to three incidents and performed normal station-level duties and training. The incident responses included an arcing power line, a food on the stove fire in a multifamily residence, and an EMS incident.

After his shift ended on the morning of March 10, 2006, Lieutenant Bowman returned home. He exercised twice during the day and had dinner with family and friends that evening. He was last seen alive while sleeping on the couch at approximately 2200hrs. At approximately 0900hrs the morning of March 11, 2006, Lieutenant Bowman was discovered deceased on the couch by his wife. The cause of death was listed as hypertensive heart disease.

For additional information regarding this incident, please refer to NIOSH Fire Fighter Fatality Investigation and Prevention Program report F2006-21 (*www.cdc.gov/niosh/fire/reports/face200621.html*).

⌒

March 12, 2006 – 1236hrs
Donald Bernard Lalosh, Firefighter
Age 49, Volunteer
Roxbury Volunteer Fire Department, New York

Firefighter Lalosh participated in the annual Saint Patrick's Day parade in Halcottsville, New York. Firefighter Lalosh carried and played a bass drum. During the parade, Firefighter Lalosh stepped out of the procession, removed his drum, and sat down. As his fire department's ambulance approached in the parade, Firefighter Lalosh began to walk toward the ambulance when he collapsed.

Firefighters and EMS workers provided assistance and Firefighter Lalosh was placed in the ambulance. Paramedics met the ambulance on the road to the hospital and continued care. Firefighter Lalosh was reported to be talking upon his arrival at the hospital but died shortly after arriving. The death was caused by a heart attack.

March 12, 2006 – 1900hrs
James Wilson McMorries, Jr., Firefighter
Age 62, Volunteer
Howardwick Volunteer Fire Department, Texas

Firefighter McMorries was operating a fire department water tender on the scene of a wildland fire. The vehicle was a converted military 6X6, 2-1/2 ton flatbed truck. A nonbaffled 1,000-gallon water tank, pump, and piping had been installed on the vehicle by members of the Howardwick Volunteer Fire Department.

The fire occurred during a period of high fire activity in the Texas panhandle. Firefighter McMorries and his crew had been on the scene of the incident for over 7 hours.

The apparatus was assigned to drive along the side of a freeway and extinguish fire. As the apparatus proceeded at a low speed, the firefighter on the front bumper signaled to Firefighter McMorries that the fire was approaching their position. As the apparatus was placed in reverse, the rear wheels lost traction in the sand and the apparatus began to roll over.

The apparatus rolled over once into a ravine, ejecting Firefighter McMorries and two other firefighters. Firefighter McMorries was not wearing a seatbelt. Firefighter McMorries received severe injuries, including a major head injury, spinal injuries, a broken back and broken ribs, and collapsed lungs. He suffered multiple CVAs during his hospital care and the decision to remove him from life support was made by his family. He died on April 9, 2006.

The Texas State Fire Marshal's Office prepared a detailed report on this incident. The report is available at *www.tdi.state.tx.us/fire/fmloddinvesti.html*

March 13, 2006 – 1140hrs
Patrick George Henry, Captain
Age 54, Wildland Full-Time
California Department of Forestry and Fire Protection

Captain Henry completed his mandatory hour-long physical fitness activity. Captain Henry's workout that day consisted of lap swimming approximately 1 mile.

After completing the workout, Captain Henry reported to the Parlin Fork Conservation Camp at 0800hrs. At 0830hrs, he assumed control of an inmate fire crew. At approximately 1030hrs, Captain Henry led his crew on a conditioning hike.

At the completion of the hike, Captain Henry returned to his office and told others in the office that his heart did not feel right. Captain Henry stood up and fell to the ground, striking his head on the desk as he fell. Others began CPR and a medical helicopter was summoned.

When helicopter paramedics arrived, a defibrillator was attached and a shock was delivered. ALS-level treatment was continued during Captain Henry's flight to the hospital. Despite these efforts, Captain Henry was not revived.

⌒

March 17, 2006 – 2345hrs
Kelly Michael Kincaid, Lieutenant
Age 41, Paid-on-Call
Morganton Department of Public Safety, North Carolina

Lieutenant Kincaid had just arrived home from a medical appointment for his wife. His pager sounded with an alarm of fire for two residences located 3 blocks from his home. Kincaid was one of the first firefighters on the scene and he assisted with extending hoselines to the rear of the residences for fire control. While working, Lieutenant Kincaid collapsed.

Firefighters found Lieutenant Kincaid without a pulse and not breathing. He was treated and transported to the hospital. Lieutenant Kincaid died as the result of his heart attack on March 19, 2006.

Lieutenant Kincaid also served as a Burke County Sheriff's Deputy. A 34-year-old man was indicted on two felony counts of first degree arson and other charges related to the fires.

⌒

March 20, 2006 – 1515hrs
Robert Eugene McLaughlin, Captain
Age 40, Career
Ocean Shores Fire Department, Washington

Captain McLaughlin and other members of the Ocean Shores Police and Fire Departments were conducting personal watercraft training at a local beach. Classroom training was completed in the morning and the class moved to the beach at approximately 1220hrs. Captain McLaughlin was wearing a helmet, a personal flotation device, and other safety equipment.

continued on next page

At approximately 1515hrs, Captain McLaughlin was a passenger in a watercraft operated by another firefighter. The watercraft was hit by a rogue wave, rolled over, and both firefighters were dumped into the water.

An instructor hovering nearby in another watercraft came to the side of Captain McLaughlin, who had drifted away from his watercraft. Captain McLaughlin advised the instructor that he was too tired to pull himself into the back of the instructor's watercraft. Captain McLaughlin was provided with a flotation device, but lost control of the device when he reached for the instructor's watercraft.

Captain McLaughlin rolled over onto his face. The instructor entered the water, found Captain McLaughlin to be unconscious, and pulled him to shore. Firefighters on the beach and an offduty nurse immediately began to provide medical care. An ambulance arrived on the scene and Captain McLaughlin was transported to the hospital. He was transported to a regional care facility by aircraft later in the day.

Captain McLaughlin died on March 22, 2006, at 0019hrs as a result of near-drowning complications.

⤳

March 12, 2006 – 2150hrs
Barry Roy Levin, Firefighter
Age 58, Volunteer
Lawrence Township Volunteer Fire Company Number 3, Pennsylvania

Firefighter Levin was driving to the fire station to perform his regular duties of checking security and preparing the furnace for the night. While driving to the station, he suffered an aortic aneurysm and was found slumped over the steering wheel at an intersection.

EMS arrived on the scene approximately 10 minutes after being called and found Firefighter Levin unresponsive. He was transported to the hospital, but was not revived.

⤳

March 21, 2006 – 1730hrs
Shon Everett Rice, Firefighter/EMT
Age 34, Career
Georgetown County Fire/EMS, South Carolina

Firefighter/EMT Rice arrived for duty at his fire station at 0745hrs. During the day, Firefighter/EMT Rice and his crew completed normal station duties. There were no incident responses during the day.

Firefighter/EMT Rice was last seen walking around the apparatus bay at approximately 1630hrs. At 1656hrs Firefighter/EMT Rice was found lying face down, unresponsive, without a pulse or respirations in the apparatus bay. CPR and ALS-level EMS care were provided in the fire station and in the ambulance enroute to the hospital. Firefighter/EMT Rice was pronounced dead at the hospital.

The cause of death was listed as "probable cardiac arrhythmia," due to "seizures," due to "epilepsy." The autopsy, completed by the county coroner's office, concluded that Firefighter/EMT Rice most likely died of a "sudden unexplained ventricular arrhythmia or possible seizure."

For additional information regarding this incident, please refer to NIOSH Fire Fighter Fatality Investigation and Prevention Program report F2006-18 (*www.cdc.gov/niosh/fire/reports/face200618.html*).

April 3, 2006 – 1000hrs
David Lewis Moore II, Assistant Chief
Age 40, Industrial
Valero Refinery Fire Brigade, Houston, Texas

Assistant Chief Moore was serving as a guest instructor at the Brayton Fire Training Field in College Station, Texas. Assistant Chief Moore was assisting with live-fire training exercises.

During rehab, Assistant Chief Moore complained of dizziness and collapsed. Other firefighters began medical care and an ambulance was called. The ambulance arrived 3 minutes after Assistant Chief Moore's collapse and began ALS-level care.

Assistant Chief Moore was transported to the hospital but was not revived. The cause of death was listed as "brain death secondary to brain aneurysm."

April 4, 2006 – Time Unknown
Richard George "Dick" Sullivan, Fire Chief
Age 57, Volunteer
Horseheads Fire Department, New York

Chief Sullivan participated in a live-fire training exercise with members of his fire department during the evening of April 4, 2006. Chief Sullivan assisted with the interior setup of the burn, deployed handlines and a supply line, assisted with pickup after the burn, and helped wash the apparatus upon its return to the fire station. The training activity ended at approximately 2230hrs.

Chief Sullivan went home and was discovered deceased the next morning. Chief Sullivan had taken a shower after arriving home, but expired prior to going to bed. The cause of death was a heart attack.

April 7, 2006 – Time Unknown
Kurt Edwin Krebbs, Captain
Age 45, Career
Oceanside Fire Department, California

Captain Krebbs fought a stubborn residential structure fire on April 7, 2006. His engine cleared the scene at 2051hrs. Captain Krebbs worked a half shift the next day and went home at approximately 1900hrs on April 8, 2006. He spoke with another firefighter on the telephone during the evening of April 8th but missed an appointment on April 9, 2006.

When he failed to report for work the next shift, firefighters went to his home to check on his welfare. They discovered Captain Krebbs deceased in his home. A medical examiner's report placed the time of Captain Krebbs' death between midnight and 0800hrs on April 9th. The death was caused by a heart attack.

April 8, 2006 - 1627hrs
Thomas James "Emmett" Kuehl, Firefighter/EMT
Age 38, Volunteer
Elkton Volunteer Fire Department, South Dakota

Firefighter/EMT Kuehl and the members of his fire department were dispatched to a controlled burn that had gone out of control. The incident was located approximately 10 miles outside of the city.

During the fire fight, two fire apparatus became stuck in a field. Firefighter/EMT Kuehl was at the wheel of one of the trucks as it was being pulled by a tractor. The tow rope disengaged from the tractor and the tension in the rope caused it to recoil toward the fire apparatus operated by Firefighter/EMT Kuehl.

Firefighter/EMT Kuehl was struck in the forehead area by a shackle attached to the tow rope. Emergency responders were called to the scene and treatment was provided to Firefighter/EMT Kuehl. He was transported by ambulance to a local hospital and subsequently transferred by medical helicopter to a regional care facility.

Firefighter/EMT Kuehl had sustained a severe head injury and died on April 11, 2006.

∽

April 11, 2006 – 0628hrs
Kevin Anthony Apuzzio, Foreman
Age 21, Volunteer
East Franklin Volunteer Fire Department, New Jersey

Foreman Apuzzio and the members of his fire department were dispatched to a residential structure fire. First-arriving police officers removed an occupant from the front porch of the structure. The occupant advised that a 75 year-old woman was trapped in the structure.

The East Franklin Fire Chief was the first firefighter on the scene and reported a working structure fire. Police officers informed the chief of the woman trapped in the house

Firefighters from the first-arriving engine company entered the structure armed with a thermal imaging camera and a charged hoseline. The firefighters located the trapped fire victim. Foreman Apuzzio and a partner arrived on the second engine. They were also equipped with a thermal imaging camera and were ordered to enter the structure and assist with the removal of the occupant.

Foreman Apuzzio and another firefighter could be seen from the outside of the structure dragging the fire victim down a hallway toward the front door. The victim was entangled in something and firefighters were having difficulty moving her. Without warning, a sudden floor collapse occurred and dropped Foreman Apuzzio, two other firefighters, and the fire victim into the basement. A large volume of fire emanated from the hole in the floor, and firefighters close to the hole were removed from the structure and moved to waiting ambulances.

Firefighters darkened down the fire in the basement and made access using an attic ladder. One firefighter was able to climb the ladder and escape the basement. Another unconscious firefighter was removed from the basement and brought into the front yard of the house for treatment by EMS responders. An accountability roll call was conducted and Foreman Apuzzio was found to be missing.

continued on next page

Foreman Apuzzio was located by the sound of his PASS device and removed from the basement by fellow firefighters. He was removed from the structure approximately 90 minutes after the collapse and transported to the hospital. He was pronounced dead shortly after arriving at the hospital.

The cause of death was listed as smoke inhalation and 40 percent body surface burns. The fire victim also was killed as a result of the incident.

April 11, 2006 – 2130hrs
David E. Smith, Fire Police Officer/Fire Commissioner
Age 74, Volunteer
Howells Fire Department, New York

Fire Police Officer responded with other members of his fire department to a hazardous materials incident at a residence. Approximately 40 gallons of diesel fuel had been spilled.

When he arrived on the scene, Fire Police Officer Smith complained of not feeling well. EMS was dispatched to the scene and Fire Police Officer Smith was evaluated and placed in the ambulance. He suffered a heart attack and could not be resuscitated.

April 13, 2006 – 1230hrs
Mark Allen Stanley, Forestry Technician
Age 49, Wildland Full-Time
Tennessee Division of Forestry

Forestry Technician Stanley and three other Tennessee Division of Forestry employees were assisting a landowner with a prescribed burn. Forestry Technician Stanley was not wearing head protection.

A bulldozer was being used to push over a standing dead tree. Forestry Technician Stanley was talking on the radio to coordinate the response to a wildland fire. As the tree fell, it struck Forestry Technician Stanley.

The dozer operator immediately went to the aid of Forestry Technician Stanley. The operator found that he was unconscious and had no vital signs. The operator called for assistance from EMS. Forestry Technician Stanley had received severe head trauma and was transported to the hospital. He was pronounced dead upon arrival.

A report on the incident cited two factors contributing to the incident. First, Forestry Technician Stanley was distracted by offsite activities and second, Forestry Technician Stanley was not at least 1-1/2 times the height of the tree from the bulldozer.

The cause of death was listed as multiple blunt force injuries.

April 15, 2006 – 0855hrs
Jose Luis Ramirez, Jr., Firefighter/EMT
Age 35, Volunteer
Far South Volunteer Fire Department, Chaparral, New Mexico

The Far South Volunteer Fire Department was paged to respond to an EMS incident involving a woman in labor. Firefighter/EMT Ramirez responded to the incident in a county-issued pickup truck.

During the response, Firefighter/EMT Ramirez was involved in a crash. As his vehicle entered a curve, it left the roadway and struck a fire hydrant and a utility pole. Firefighter/EMT Ramirez was not wearing a seatbelt.

Firefighters responded to the crash site and extricated Firefighter/EMT Ramirez. He was found to be pulseless. Medical treatment, including CPR, was provided prior to the arrival of paramedics. CPR was discontinued after approximately 30 minutes and after consultation with medical control. Firefighter/EMT Ramirez was pronounced dead at the scene.

⸺

April 16, 2006 – 2200hrs
Russell Blanton Schwantes, Fire Apparatus Operator
Age 39, Career
Atlanta Fire-Rescue Department, Georgia

Fire Apparatus Operator Schwantes was on duty at an airport fire station. He had reported for duty at approximately 0700hrs and performed normal station duties during the shift.

At approximately 2100hrs, he worked out on a treadmill. At approximately 2200hrs, units from his station were dispatched to an incident. Fire Apparatus Operator Schwantes ran to the apparatus floor. When he realized that the call was not for his unit, he returned to the exercise room. Approximately 5 minutes later, Fire Apparatus Operator Schwantes entered the day room, said that he needed help, and collapsed.

Firefighters provided assistance, and an ALS ambulance arrived at the station. Fire Apparatus Operator Schwantes was assessed and found to be suffering from an acute heart attack. He was transported to the hospital and subsequently underwent cardiac catheterization and the placement of a stent.

Fire Apparatus Operator Schwantes remained in intensive care for 9 days prior to his death on April 25, 2006.

For additional information regarding this incident, please refer to NIOSH Fire Fighter Fatality Investigation and Prevention Program report F2006-20 (*www.cdc.gov/niosh/fire/reports/face200620.html*).

⸺

April 16, 2006 – 1945hrs
Willie Macon Price, Firefighter
Age 58, Volunteer
Jamesville Community Volunteer Fire Department, North Carolina

Firefighter Price drove one of his fire department's engines to a standby assignment at a fireworks event. As he stood near the truck and talked with bystanders, he suddenly collapsed of a heart attack.

continued on next page

A bystander provided CPR and an ambulance that was also standing by at the event was summoned. Firefighter Price was transported to the hospital by ambulance. He was pronounced dead at the hospital.

~

April 19, 2006 – 1800hrs
Garry Tanner, Assistant Chief
Age 57, Volunteer
Pleasantville Volunteer Fire Department, Tennessee

Assistant Chief Tanner drove a fire department water tender to the scene of a reported structure fire. The fire was found to involve a clothes dryer, and fire department members were completing their work on the scene.

Assistant Chief Tanner told other firefighters that he was going to refill the water tank on the apparatus and refuel the vehicle. As he drove into town, Assistant Chief Tanner was followed by another firefighter.

The firefighter following the water tender observed that Assistant Chief Tanner was having some sort of mechanical problem with the vehicle. Prior to entering the town, the apparatus would have to proceed down a steep hill. Rather than enter town with a malfunctioning apparatus, Assistant Chief Tanner chose to drive the apparatus into a gravel pit.

As the apparatus entered the pit area down a dirt ramp, Assistant Chief Tanner lost control of the water tender. The vehicle rolled several times and Assistant Chief Tanner was ejected through the windshield. Assistant Chief Tanner was not wearing a seatbelt. He received a fatal head injury.

An inspection of the apparatus after the crash found a cracked and leaking brake line. The Tennessee Highway Patrol inspector concluded that the loss of brake fluid would lead to brake failure.

~

April 21, 2006 – 2130hrs
William Lewis Robinson III, Fire Chief
Age 39, Volunteer
Sarge Creek Volunteer Fire Department, Oklahoma

Chief Robinson was participating in a controlled burn of pasture land. He was operating a 6X6 2-1/2 ton military surplus truck that had been converted for firefighting. The water tank on the apparatus had recently been filled.

Based on the law enforcement report on the incident, Chief Robinson dismounted the apparatus for an unknown reason and was run over by the left front wheel of the vehicle. Chief Robinson was able to move himself out of the path of the apparatus before the rear wheel passed.

His injury was discovered when the apparatus he had been driving rolled forward at a low speed and struck another piece of fire apparatus on the scene. The parking brake of Chief Robinson's apparatus was non-functional.

Chief Robinson was transported to a local hospital, but died of traumatic injuries. A police report on the incident indicated that Chief Robinson was intoxicated at the time of the incident.

∽

April 22, 2006 – 0130hrs
Joseph Louis Bilka, Firefighter
Age 60, Volunteer
Antioch-Fishing Creek Volunteer Fire Department, North Carolina

Firefighter Bilka was responding to his fire station in his personal vehicle after being dispatched to a smoke investigation. The weather conditions were foggy, and visibility was poor.

During the response, Firefighter Bilka rounded a curve and struck a flatbed trailer that was blocking the roadway. Firefighter Bilka was trapped in his vehicle underneath the trailer. Firefighters responding to reports of the crash extricated Firefighter Bilka. He was transported by medical helicopter to a regional medical facility.

Firefighter Bilka was hospitalized until his death on May 21, 2006. The cause of death was listed as blunt force trauma.

∽

April 30, 2006 – 1644hrs
Alan Dale Leake, Captain
Age 51, Career
City of Fulton Fire Department, Missouri

Captain Leake and his engine crew were on the scene of a three-vehicle motor vehicle crash. It had been raining heavily and when the rain stopped, the weather conditions were hot and humid. As firefighters were clearing the scene of the crash, a law enforcement officer alerted them to a serious crash nearby.

Captain Leake and his crew arrived at the scene of the second crash. This incident involved the ejection of one of the drivers. Law enforcement officers were performing CPR on the driver and Captain Leake and a firefighter climbed down an embankment to assist. Paramedics arrived, the driver was placed on a backboard, and Captain Leake assisted with carrying the driver up the embankment to the ambulance.

After the patient was transported, Captain Leake stood at the roadside and talked with the fire chief. He told the fire chief that he was feeling dizzy and then suddenly collapsed. Firefighters provided medical care, and another ambulance was called to the scene. Firefighters performed CPR until the arrival of the ambulance. Captain Leake was transported to the hospital but did not survive the incident. His death was caused by a heart attack.

∽

May 4, 2006 – 0120hrs
William LeRoy Craddock II, Firefighter III
Age 37, Career
DeKalb County Fire Rescue Department, Georgia

Prior to reporting for duty at 1800hrs on May 3, 2006, Firefighter Craddock conducted and participated in an 8-hour live-fire training exercise. Just after midnight on May 4, 2006, Firefighter Craddock and members of his fire department were dispatched to a residential structure fire. Firefighter Craddock was assigned to a heavy rescue company.

continued on next page

As firefighters arrived on the scene, they found a working fire in a trilevel residence. Firefighter Craddock and his crew advanced an attack line into the structure to engage the fire. The crew breached walls to apply water to the fire and to check for fire spread. Firefighter Craddock's low-air alarm began to sound and he exited the structure.

When Firefighter Craddock emerged from the structure, he and other firefighters removed his protective clothing and his SCBA. Firefighter Craddock's company officer noticed that Firefighter Craddock did not look well and had him evaluated by paramedics. Despite an improvement in his condition during treatment, a command officer ordered Firefighter Craddock to be transported to the hospital for evaluation.

Upon his arrival at the hospital, he began to exhibit signs of a heart attack. His condition worsened and progressively more aggressive life-saving measures were used. Firefighter Craddock was transferred by ambulance to a regional care facility. His condition worsened during the transport. Firefighter Craddock survived his heart attack but remained in critical condition for 39 days. He suffered a number of complications and was removed from life support on June 13, 2006. He died shortly thereafter.

For additional information regarding this incident, please refer to NIOSH Fire Fighter Fatality Investigation and Prevention Program report F2006-17 (*www.cdc.gov/niosh/fire/reports/face200617.html*).

⟊

May 14, 2006 – 0430hrs
Richard Patrick "Rich" Montoya, Lieutenant
Age 61, Career
Denver Fire Department, Colorado

Lieutenant Montoya's engine company, along with other Denver Fire Department units, was dispatched to a report of a structure fire in a residence. The caller reported that one person was trapped in the structure.

Firefighters arrived on the scene and found a working fire in a two-story structure. Firefighters entered the house to perform a search; they located a victim and removed her from the structure. Lieutenant Montoya's engine company laid a supply line from a hydrant and advanced an attack line into the structure. Firefighters advanced the attack line to the second story. Lieutenant Montoya, who had been on the nozzle, gave the nozzle to his firefighter.

Smoke and heat conditions on the second floor began to worsen. Firefighters had difficulty in finding the fire. The ceiling was opened and water was applied to the attic. Lieutenant Montoya's firefighter communicated with Lieutenant Montoya that the crew should go back to the stairs to regroup. Thinking Lieutenant Montoya had exited before him, the firefighter left the structure.

Firefighters operating inside of the structure heard the faint sound of a PASS device and began a search. Despite difficult fire and debris conditions, firefighters found Lieutenant Montoya unconscious under a mattress. Firefighters reported to the Incident Commander (IC) that a firefighter was down; Command activated the Rapid Intervention Team (RIT).

Fire conditions were worsening, and firefighters crawled on their stomachs to push and drag Lieutenant Montoya to the stairs. Additional firefighters and RIT members removed Lieutenant Montoya down the stairs and from the building. The firefighting strategy was changed to defensive after his removal.

continued on next page

Lieutenant Montoya was found to be in full cardiac arrest. CPR was initiated and paramedic-level EMS care was provided. Sometime prior to his arrival at the hospital, a pulse was restored. Upon his arrival at the hospital, Lieutenant Montoya's carboxyhemoglobin level was 23 percent.

Lieutenant Montoya remained in intensive care for 7 days. With no prognosis for improvement, life support was removed and he died on May 21, 2006. The cause of death was oxygen deprivation to the brain as a result of smoke inhalation. Lieutenant Montoya was 15 shifts away from his planned retirement.

⌒

May 21, 2006 – 1429hrs
Gerald A. Machajewski, Firefighter
Age 62, Volunteer
Cambria Volunteer Fire Company, New York

Firefighter Machajewski and members of his fire department responded to a vehicle crash. Once on the scene, Firefighter Machajewski assisted with patient treatment by transporting equipment from ambulances to the treatment area. He made multiple trips and carried backboards, a gurney, and other equipment. While he worked, Firefighter Machajewski was wearing full structural firefighting protective clothing.

Two children were involved in the crash incident. Firefighter Machajewski was directed to supervise the children in the back of an ambulance as treatment of others injured in the crash continued.

After the children entered the ambulance, Firefighter Machajewski entered and closed the door. Immediately thereafter, Firefighter Machajewski collapsed and became unconscious. One of the children opened a door and called for help.

Firefighters and other emergency workers began CPR and Firefighter Machajewski was transported to the hospital. He was pronounced dead at the hospital due to a heart attack.

⌒

June 13, 2006 – 2121hrs
Michael James Day, Deputy Assistant Chief
Age 49, Career
Providence Fire Department, Rhode Island

Deputy Assistant Chief Day began his regularly assigned shift at 1645hrs. He responded to a working fire in a residence and was on scene for approximately 35 minutes. At the conclusion of the incident, Deputy Assistant Chief Day returned to his quarters and performed administrative duties. He complained to other firefighters of feeling hot, sweaty, and uncomfortable.

The air conditioning in the fire station had failed, and firefighters set up a fan to provide some relief in Deputy Assistant Chief Day's office. He declined to eat dinner with the crews assigned to the station, saying that he did not feel well.

A battalion chief arrived to attend a pre-arranged meeting with Deputy Assistant Chief Day. When the battalion chief entered the office, he found Chief Day unconscious on the floor. CPR was initiated by firefighters and an AED was applied. An arriving rescue company provided paramedic-level EMS care. Deputy Assistant Chief Day was transported to the hospital.

continued on next page

Despite the efforts of firefighters and emergency room staff, Deputy Assistant Chief Day was not revived and he was pronounced dead at 2135hrs. The cause of death was listed as atherosclerotic and hypertensive cardiovascular disease.

⌾

June 16, 2006 – 1945hrs
Michael J. Greene, Firefighter
Age 43, Volunteer
West Babylon Fire Department, New York

Firefighters responded to a structure fire in a restaurant on June 13, 2006. During the course of the fire fight, a ventilation hole was cut into an awning. At the conclusion of the incident, a tarpaulin was placed over the hole by firefighters.

On Friday, June 16, 2006, Firefighter Greene and other firefighters went to the restaurant to retrieve the tarp. Firefighter Greene climbed a ground ladder to access the tarp; he was at the tip of the ladder when he made contact with an electrified sign on the awning and was electrocuted. Firefighter Green was rendered unconscious immediately and was hanging upside down from the ladder.

Firefighter Greene was brought down from the ladder, treated, and transported to a hospital. He did not survive. The sign was found to be improperly grounded.

⌾

June 23, 2006 – 2355hrs
Robert Wallace "Bobby" Edwards, Firefighter
Age 45, Volunteer
Tunica Volunteer Fire Department, Mississippi

Firefighter Edwards and the members of his department responded to the report of a fire in a manufactured home. When firefighters arrived on the scene at 0508hrs, they found the home completely involved in fire. Since there was no action that could be taken, the IC released all but the first-arriving fire company. Firefighter Edwards returned to quarters without having arrived on the scene of the incident.

At approximately 2355hrs that evening, Firefighter Edwards suffered a fatal heart attack at his residence. The cause of death was listed as hypertensive heart disease and coronary artery disease.

⌾

June 25, 2006 – 1432hrs
Steven A. Smith, Deputy Chief
Age 34, Volunteer
Wea Township Volunteer Fire Department, Indiana

Deputy Chief Smith and the members of his fire department were dispatched to a report of a structure fire. Lightning had struck a house and ignited a fire in the basement.

continued on next page

Deputy Chief Smith was the first firefighter on the scene of the incident and found a working fire. Wearing full structural firefighting protective clothing and an SCBA, Deputy Chief Smith and another firefighter entered the structure and found the floor to be spongy. Deputy Chief Smith ordered the other firefighter to apply water to the basement through a side window. Deputy Chief Smith advanced an attack line through the front door of the residence. Immediately upon re-entering the structure, Deputy Chief Smith fell into the basement through the fire-weakened floor.

Other firefighters arriving on the scene found the attack line leading into the basement. Deputy Chief Smith called to firefighters from the basement and told them that he was trapped and in need of rescue. Firefighters attempted to enter the basement through the hole using a ground ladder, but were unable to make access.

The homeowner directed firefighters to the basement stairway, and two firefighters entered the basement to search for Deputy Chief Smith. Firefighters followed the sound of Deputy Chief Smith's PASS device and located him; he was unconscious and his facepiece was not in place. Firefighters began to drag Deputy Chief Smith toward the stairs. Deputy Chief Smith became entangled in debris, and firefighters attempted to remove his SCBA and provide air from their SCBAs.

Deputy Chief Smith was removed from the building and transported to the hospital by ambulance. He was pronounced dead after his arrival. The cause of death was listed as smoke inhalation. At autopsy, Deputy Chief Smith's carboxyhemoglobin level was 57 percent.

The flooring system that collapsed was an engineered product consisting of solid wood top and bottom plates and a chipboard web. This type of truss has been involved in several firefighter fatalities and is the subject of a Special Topic section in this report.

The Wea Township Fire Department was cited by the Indiana Department of Labor for five serious violations of standards including facepiece fit testing, lack of medical evaluations for firefighters, and two-in/two-out protocols.

July 4, 2006 – 0832hrs
Eric James Olson, Firefighter
Age 35, Volunteer
Laury's Station Volunteer Fire Company Number One, Pennsylvania

Firefighter Olson and the members of his fire company responded to a mutual-aid river rescue on July 4, 2006. Firefighter Olson arrived at the scene at 1431hrs as the driver of an attack/brush unit.

While on the scene of the incident, Firefighter Olson handed out river rescue gear and remained with the unit in staging. Firefighters from Firefighter Olson's fire company stood by on the river banks with rope throw bags as a part of a rescue effort upstream from their position. The incident was concluded by 1600hrs.

Firefighter Olson collapsed at home as the result of a heart attack at 0832hrs on July 5, 2006. He was transported to the hospital but did not survive.

July 12, 2006 – 0735hrs
Otis Wayne Chupp, Fire Chief
Age 61, Career
Cartersville Fire Department, Georgia

At approximately 0650hrs, Chief Chupp met another fire department member at a local park for an annual physical fitness certification. The certification involved walking or running 1-1/2 miles on a quarter-mile track. Chief Chupp completed his certification in 23 minutes and 43 seconds. Chief Chupp cooled off on the track and left for home.

At 1136hrs an emergency call was received from Chief Chupp's residence. Chief Chupp's wife had discovered him unconscious in the carport between two vehicles. Chief Chupp was dressed in the same clothes that he wore for his certification test and was obviously deceased. Apparently, he collapsed upon his return home and never made it into the house.

The cause of death was listed as a heart attack.

⟿

July 26, 2006 – 1915hrs
Alethea Faye Nixon, Junior Firefighter
Age 17, Volunteer
Asbury Volunteer Fire Department, Alabama

Firefighter Nixon was a passenger in a 2,000-gallon fire department water tender responding to a report of a structure fire in a house. Including the driver, there were four occupants in the cab of the apparatus. The cab of the apparatus was designed for two occupants.

As the water tender approached a single-lane bridge, the driver applied the brakes but was unable to slow the apparatus. The apparatus crashed through the bridge railing and plunged into a creek bed 30 feet below. The apparatus came to rest on its roof. The driver was able to self-extricate and was climbing the bank of the creek to seek assistance when other firefighters arrived.

Arriving firefighters and rescue personnel extricated the three injured firefighters. Two were transported to hospitals in critical condition, and Junior Firefighter Nixon was pronounced dead at the scene. The cause of death was multiple impact trauma.

For additional information regarding this incident, please refer to NIOSH Fire Fighter Fatality Investigation and Prevention Program report F2006-25 (*www.cdc.gov/niosh/fire/reports/face200625.html*).

⟿

July 27, 2006 – 1135hrs
Christopher Ryan Brown, Firefighter
Age 20, Volunteer
LaHarpe Volunteer Fire Department, Kansas

Firefighter Brown was on a work detail driving from his town to Manhattan, Kansas, to pick up fire equipment that was being given to his department by the Kansas State Forestry Department. Firefighter Brown was driving his personal vehicle, a 2001 Dodge pickup. A police report on the incident noted that it was raining.

continued on next page

For reasons unknown, Firefighter Brown's vehicle crossed the centerline of the highway and collided head-on with a tractor-trailer truck headed in the opposite direction. Both vehicles were traveling at highway speeds. Firefighter Brown sustained major injuries and he was pronounced dead at the scene. Firefighter Brown was wearing his seatbelt at the time of the crash.

✑

July 28, 2006 – 0844hrs
Rose Ann Woodbridge, Firefighter
Age 56, Volunteer
Hanover Township Fire Department, Ohio

Firefighter Woodbridge was responding in her personal vehicle to a motor vehicle crash. She was responding with lights and siren in operation. The roads were wet from recent rainfall.

Firefighter Woodbridge entered a curve in the roadway, went left of center, and overcorrected. The car began to slide sideways, left the roadway, and then crashed into a utility pole. The passenger door of the car hit the pole, and the vehicle sustained heavy damage.

Firefighter Woodbridge was extricated from the vehicle and transported by ambulance to a landing site where responders met a medical helicopter. A physician arriving with the medical helicopter pronounced her dead after assessing her injuries. She had sustained numerous injuries in the crash, including a serious head injury.

Firefighter Woodbridge was wearing her seatbelt at the time of the crash.

✑

July 29, 2006 – 2227hrs
Jeffrey Dean Irwin, Firefighter
Age 43, Paid-on-Call
Eldorado Fire Department, Illinois

Firefighter Irwin and the members of his department responded to a structure fire incident at 1917hrs on July 29th. The heat index was above 100 °C (38 °F) . Firefighter Irwin assisted with extending attack lines and setting up equipment at the scene of the incident . Three firefighters, including Firefighter Irwin, became ill at the scene of the incident due to overexertion and overheating.

At the conclusion of the incident, Firefighter Irwin returned to his home at 2100hrs. Firefighters called and stopped by to check on his status several times. Firefighter Irwin was found deceased at 2227hrs by a firefighter's wife who stopped to check on him. The firefighter's wife called 9-1-1 and initiated CPR. When the ambulance arrived, Firefighter Irwin was pronounced dead.

The autopsy concluded that the cause of death was atherosclerotic cardiovascular disease.

✑

July 30, 2006 – 1010hrs
Paul R. Brady, Firefighter
Age 42, Volunteer
Malverne Fire Department, New York

Firefighter Brady and other firefighters were performing routine maintenance at the Malverne fire station. Firefighter Brady was on top of the department's heavy rescue truck checking on supplies that are stored in that area of the truck.

Not knowing that Firefighter Brady was on top of the apparatus, a firefighter began to move the rescue truck outside to allow the floor below to be cleaned. As the vehicle began to move, an officer realized that Firefighter Brady was on top of the vehicle and called to the driver. The apparatus was placed in reverse and Firefighter Brady fell from the roof of the rescue to the floor. Firefighter Brady was severely injured after being crushed between the door frame and the apparatus.

Firefighter Brady was transported to the hospital, but was pronounced dead at approximately 1411hrs. The death was caused by traumatic injuries.

⟶

August 2, 2006 – 0200hrs
Larry Wayne Fanning, Firefighter
Age 57, Volunteer
Garner Fire Department, North Carolina

Firefighter Fanning was at the fire station conducting fire department business when the department received a call for a medical emergency. Firefighter Fanning did not respond on the incident but stood by at the station.

At the conclusion of the incident, Firefighter Fanning returned to his home where he died of a heart attack a few hours later.

⟶

August 2, 2006 – 0620hrs
Lee William Walters, Assistant Chief
Age 54, Part-Time (Paid)
Sheldon Fire District, South Carolina

Assistant Chief Walters and the members of his fire department were dispatched to a report of a fire in a residence. Upon their arrival, firefighters found a well-involved single-story wood-frame structure. Firefighters established a water supply and fought the fire with multiple handlines.

About 20 minutes into the operation, Assistant Chief Walters suddenly collapsed. Emergency medical care was provided immediately, and Assistant Chief Walters was transported to a hospital. He was not revived and was pronounced dead at the hospital. The cause of death was listed as a heart attack.

⟶

August 4, 2006 – 1947hrs
Terry Wayne Jacobs, Pilot
Age 58, Wildland Contract
Heavy Lift Helicopters under contract to the USDA Forest Service, California

Andrei Pantchenko, III, Co-Pilot
Age 38, Wildland Contract
Heavy Lift Helicopters under contract to the USDA Forest Service, California

Pilot Jacobs and Co-Pilot Pantchenko were operating a Sikorsky CH-54A helicopter in support of fire-fighting operations at the Titus fire near Happy Camp, California.

The helicopter had been returned to service earlier in the day after having an engine replaced. After completing a series of tests, the helicopter was assigned to the Titus fire and conducted a number of water drops. After over 2 hours of work, the helicopter returned to its helibase for fueling. The unit departed base at 1912hrs for another round of water drops.

As the helicopter maneuvered near a water dip site, it experienced an in-flight separation of a tail rotor blade. A witness heard a bang, saw pieces fall from the rear of the helicopter, and observed a crash. Both crew members were killed.

The cause of death for Pilot Jacobs was listed as multiple trauma and the cause of death for Co-Pilot Pantchenko was listed as drowning.

For additional information about this crash, consult the National Transportation Safety Board Web site at *www.ntsb.gov/ntsb/query.asp* and use NTSB identification LAX06GA254.

August 8, 2006 – 1000hrs
Richard Washington, Jr., Firefighter II
Age 49, Career
Clark County Fire Department, Nevada

Firefighter Washington was performing onduty physical fitness activities. He had been on a treadmill for approximately 20 minutes when he collapsed.

Another firefighter in the area heard a change in the sound of the treadmill and came to check on Firefighter Washington. The firefighter found him on top of the treadmill and unresponsive. The firefighter called for assistance.

ALS-level emergency medical care was provided at the fire station and while enroute to the hospital, but Firefighter Washington was pronounced dead at the hospital at 1030hrs.

After an autopsy was completed, the cause of death was listed as atherosclerotic cardiovascular disease.

August 8, 2006 – 2015hrs
Ollie Phil Tate, Firefighter
Age 56, Volunteer
Thaxton Volunteer Fire Department, Mississippi

Firefighter Tate was a passenger in the bed of a pickup truck that was responding to a confirmed structure fire. He suffered a heart attack and fell from the bed of the pickup.

∽

August 10, 2006 – 2259hrs
Dana Lynn MacCrimmon, Firefighter
Age 45, Career
Carbondale Fire Department, Illinois

Firefighter MacCrimmon returned from an emergency response to a residential board and care facility. The fire alarm was triggered by smoke produced by cooking. Firefighters reset the alarm and returned to quarters.

Upon her arrival at the fire station, Firefighter MacCrimmon complained of head pain and asked to be taken to the hospital. Firefighters transported her in a fire department car; by the time they arrived at the hospital, Firefighter MacCrimmon was unresponsive. She was treated at the hospital but her condition worsened and she was pronounced dead at 1420hrs on August 11, 2006.

The cause of death was reported to be a CVA (stroke).

∽

August 12, 2006 – 0642hrs
Dennis K. "Denny" Hayes, Firefighter
Age 59, Volunteer
McClure Fire Company, Pennsylvania

Firefighter Hayes was operating the pump panel on an engine at the scene of a mutual-aid structure fire. He suddenly collapsed. He was treated by paramedics on the scene and transported to a local hospital. He was pronounced dead at the hospital as the result of a heart attack.

∽

August 13, 2006 – 1223hrs
Arnold W. "Arnie" Wolff, Lieutenant
Age 55, Career
Green Bay Fire Department, Wisconsin

Lieutenant Wolff was assigned to an ambulance on August 13, 2006. The ambulance was dispatched, along with other fire department units, to a structure fire in a residence. Firefighters arriving on the scene found a working fire with smoke showing.

Lieutenant Wolff and an engineer were ordered to enter the front door of the house and perform a left-hand search. Another two-firefighter crew entered and went to the right. Within minutes of their entry, a partial

continued on next page

floor collapse occurred, and Lieutenant Wolff and the engineer fell approximately 10 feet into the fully-involved basement. Maydays were transmitted by Lieutenant Wolff and the engineer.

A second alarm was immediately requested by the IC and firefighters began rescue efforts. The engineer fell into a finished portion of the basement that had windows. She was assisted through a window by other firefighters.

Lieutenant Wolff fell into a room that did not have windows, and his path to an exit was blocked by debris. Fire conditions advanced markedly after the collapse, and firefighters were unable to reach Lieutenant Wolff. His body was recovered approximately 13 hours into the incident.

The cause of death was listed as asphyxiation due to smoke inhalation. At autopsy, the carboxyhemoglobin level in Lieutenant Wolff's blood was 49.6 percent. The fire cause is undetermined but it was concluded that the fire burned for some time in the basement prior to its discovery. The trusses that failed were in the area that was most damaged by fire.

The trusses that failed were TGI trusses, an engineered lumber product. Additional information on engineered lumber is contained in the Special Topic section of this report.

⌒

August 13, 2006 – 0310hrs
Paul Raymond Montavon, Sr., Lieutenant
Age 59, Part-Time (Paid)
Whitewater Township Fire Department, Ohio

Lieutenant Montavon and members of his fire department were on the scene of a motor vehicle crash involving a pedestrian. As Lieutenant Montavon assisted at the scene, he collapsed.

Lieutenant Montavon was placed into an ambulance and found to be in cardiac and respiratory arrest. He was transported to the hospital where he was pronounced dead. His death is likely due to a heart attack.

⌒

August 13, 2006 – 1756hrs
Quinn Russell Stone, Pilot
Age 42, Wildland Contract
Evergreen Helicopters under contract to the USDA Forest Service, Yellow Pine, Idaho

Michael Gene Lewis, Firefighter/Assistant Helitack Manager
Age 37, Wildland Full-Time
USDA Forest Service - Payette National Forest, Idaho

Lillian M. Patten, Firefighter/Lookout
Age 32, Wildland Full-Time
USDA Forest Service - Payette National Forest, Idaho

Monica Lee Zajanc, Firefighter
Age 27, Wildland Part-Time
USDA Forest Service - Payette National Forest, Idaho

continued on next page

These firefighters were aboard a Eurocopter AS-350-B3 helicopter. The helicopter was in the middle of a mission to replace the crew staffing a fire watch station on the top of Williams Peak.

The drop was completed and the offgoing crew was aboard the helicopter returning to the base. For reasons unknown, the helicopter crashed on a mountainside at an elevation of approximately 1,800 feet. All aboard were killed.

For additional information about this crash, consult the National Transportation Safety Board Web site at *www.ntsb.gov/ntsb/query.asp* and use NTSB identification SEA06GA158.

∾

August 17, 2006 – 1350hrs
Spencer Stanley Koyle, Division Supervisor
Age 33, Wildland Full-Time
Bureau of Land Management, Fillmore Field Office, Utah

Division Supervisor Koyle was on the scene of the lightning-caused Devil's Den wildland fire near Delta, Utah. He was assigned as the Assistant Fire Management Officer for this incident.

Division Supervisor Koyle and the IC were airlifted into the area of the fire at approximately 1230hrs. After his arrival, Division Supervisor Koyle walked down into the canyon to scout the fire. After working his way down into the canyon, Division Supervisor Koyle ordered water drops on hot spots.

Fire conditions worsened dramatically, and the IC ordered Division Supervisor Koyle to get out of the canyon. Division Supervisor Koyle began to run from the advancing fire, stopped to deploy his fire shelter, and was overrun by the fire. The fire shelter was not able to protect him sufficiently, and he was fatally burned.

An accident investigation report on the incident cited four causal factors in the fatality: 1) Division Supervisor Koyle ignored the IC's advice to stay out of the canyon; 2) Division Supervisor Koyle lost awareness of the fire's blowup potential; 3) Division Supervisor Koyle lost his situational awareness as he focused on directing water drops; and 4) Division Supervisor Koyle violated a number of standard fire-fighting orders.

∾

August 27, 2006 – 1330hrs
Howard John Carpluk, Jr., Lieutenant
Age 43, Career
Fire Department City of New York, New York

Michael C. Reilly, Firefighter
Age 25, Career
Fire Department City of New York, New York

Lieutenant Carpluk (assigned to Engine 42 and working overtime at Engine 75) and Firefighter Reilly were working at Engine Company 75 on August 27, 2006. The company was dispatched as an additional unit to a working fire in a convenience store. Lieutenant Carpluk, Firefighter Reilly, and other firefighters advanced a second 2-1/2-inch attack line into the structure to support extinguishment operations.

continued on next page

A collapse occurred involving an area of the first floor of the structure. Lieutenant Carpluk and Firefighter Reilly fell into the basement. Lieutenant Carpluk was entrapped by the collapse debris, and Firefighter Reilly was encapsulated in the collapse debris. Other firefighters were caught in the collapse area as well. Firefighters were able to remove the members trapped in the upper part of the collapse area. It took an hour to remove the last of these members.

During that time, other members breached holes into the collapse area from an adjoining cellar. These members found Lieutenant Carpluk approximately 15 minutes after entering the collapse area. Lieutenant Carpluk was conscious when reached and he told firefighters that his nozzleman, Firefighter Reilly, was trapped under him. The extrication of Lieutenant Carpluk took more than 45 minutes. He was in respiratory arrest when he was removed from the structure. One to 2 minutes after Lieutenant Carpluk was removed, rescuers found Firefighter Reilly in the collapse debris. It took approximately 20 minutes to extricate Firefighter Reilly.

Firefighter Reilly died from asphyxia due to compression of the chest. Lieutenant Carpluk died of positional asphyxia and aspiration of blood.

The structure that was involved in fire had experienced a major-alarm fire in 2000.

∽

August 28, 2006 – 0946hrs
Wilbur A. Ritter, Fire Police Officer
Age 78, Volunteer
Sayville Fire Department, New York

Fire Police Officer Ritter and the members of his fire department were dispatched to an alarm at a local drug store for a hazardous condition with the potential for a roof collapse.

Fire Police Officer Ritter arrived at the fire station and prepared to respond to the scene in the department's fire police vehicle. He was observed to be looking ill and complained of chest pains and difficulty breathing.

Firefighters summoned EMS responders and provided Fire Police Officer Ritter with oxygen. He was transported to the hospital, where he later died.

The cause of death was listed as atherosclerotic heart disease.

∽

September 4, 2006 – 1353hrs
Errett Wayne Miller, Assistant Fire Chief
Age 43, Volunteer
Posey Township Volunteer Fire Department, Indiana

Assistant Chief Miller was the driver and sole occupant of a 1,500-gallon water tender responding to an arson-caused mutual-aid structure fire.

The water tender entered a 20-degree left-hand curve at a speed estimated at 63 mph. The posted speed limit for the roadway was 55 mph. As the apparatus negotiated the curve, the right wheels of the water

continued on next page

tender left the paved surface of the road and the vehicle traveled approximately 141 feet on the grassy shoulder. Assistant Chief Miller overcorrected and brought the water tender back onto the roadway.

The water tender began rotating counter-clockwise, Assistant Chief Miller again overcorrected, the water in the apparatus tank shifted, and the apparatus began to rotate clockwise. The apparatus eventually left the roadway and rolled four times, coming to rest on the driver's side. Between the second and third rolls of the apparatus, Assistant Chief Miller was ejected.

Assistant Chief Miller received mortal injuries. Responders provided emergency medical care and he was transported to a hospital. Assistant Chief Miller was pronounced dead on arrival at the hospital.

A law enforcement report on the incident cited speed as a factor in the crash. The report observed that Assistant Chief Miller did everything that he could to recover control of the vehicle, but was not able to overcome the speed of the vehicle and the instability of the water.

The cause of death was listed as head and chest trauma.

∽

September 4, 2006 – 1400hrs
David F. Prior, Lieutenant
Age 50, Volunteer
Elbridge Volunteer Fire Company, Inc., New York

Lieutenant Prior and the members of his fire department participated in the Labor Day parade and push ball competition in Skaneateles, New York. Lieutenant Prior marched in the parade which lasted approximately 30 to 45 minutes.

At the completion of the parade, Lieutenant Prior complained of the heat, said that he was having difficulty breathing, and said that he was experiencing indigestion. As firefighters prepared for the pushball competition, Lieutenant Prior suddenly collapsed.

Firefighters and other emergency responders immediately provided assistance. CPR was initiated, and Lieutenant Prior was transported to the hospital by ambulance.

Approximately an hour after his collapse, Lieutenant Prior was pronounced dead at the hospital. The cause of death was listed as atherosclerotic and hypertensive cardiovascular disease.

September 6, 2006 – 1025hrs
Robert Paul Stone, Battalion Chief
Age 36, Career
California Department of Forestry and Fire Protection

George Elvin "Sandy" Willett, Jr., Pilot
Age 52, Wildland Contract
DynCorp under contract to the California Department of Forestry and Fire Protection

Pilot Willett and Battalion Chief Stone were assigned aerial observation duties for the "Mountain Incident" wildland fire that was in progress within the Mountain Home Demonstration State Forest near Porterville, California. The firefighters were operating a North American OV-10A aircraft.

continued on next page

The crew departed the airport normally and conducted routine communications with the command center. The aircraft was seen flying between 400 and 600 feet above the treetops by an observer.

For reasons unknown, the aircraft hit the tree tops in a box-like canyon. The initial impact was to the top of trees estimated to be 125 feet tall. The plane broke up, fell to the forest floor, and caught fire. Both occupants were killed.

For additional information about this crash, consult the National Transportation Safety Board Web site at *www.ntsb.gov/ntsb/query.asp* and use NTSB identification LAX06GA287.

∼

September 9, 2006 – 0517hrs
Vincent R. Neglia, Acting Captain
Age 45, Career
North Hudson Regional Fire & Rescue Department, New Jersey

Firefighter Neglia and other firefighters were dispatched to a report of fire in a three-story apartment building in Union City. Upon their arrival at the scene, firefighters found light smoke and no visible fire.

Based on reports that the structure had not been evacuated, Firefighter Neglia and other firefighters entered the building to perform a search. Due to the light smoke conditions, Firefighter Neglia was not wearing his facepiece.

Firefighter Neglia was the first firefighter to enter an apartment. Conditions deteriorated rapidly as fire broke through a ceiling and caused a collapse. Firefighter Neglia was trapped by the collapse and rapid fire progress. Other firefighters came to his aid and removed him from the building. Firefighter Neglia was transported to the hospital but later died of a combination of smoke inhalation and burns.

∼

September 16, 2006 – 1316hrs
John Paul "JP" Memory II, Firefighter
Age 19, Volunteer
Carrollton Volunteer Fire Department, Virginia

Firefighter Memory and other firefighters participated in a Patriot Day parade in the community of Franklin, Virginia. Firefighter Memory and his rescue company were asked by Franklin firefighters to participate in an extrication demonstration at the conclusion of the parade.

Firefighter Memory was operating a hydraulic rescue tool as a part of the demonstration. He told other firefighters that he was not feeling well, passed the tool to another firefighter, and went to sit down on the back of the apparatus. Moments later, firefighters found Firefighter Memory collapsed on the ground.

CPR and paramedic-level emergency medical care were initiated immediately. Firefighter Memory was transported to the hospital by ambulance. He was pronounced dead at the hospital.

Firefighter Memory's death was caused by dilated cardiomyopathy.

∼

September 18, 2006 – 1526hrs
Johnny Nelson Buchanan, Lieutenant
Age 47, Volunteer
Feds Creek/Mouth Card Area Volunteer Fire Department, Kentucky

Lieutenant Buchanan and the members of his fire department responded to an emergency medical incident on September 17, 2006. The incident occurred at 1827hrs and was concluded shortly thereafter.

On September 18, 2006, at 1526hrs, Lieutenant Buchanan suffered a heart attack while driving. His vehicle crossed the center line of the roadway and slowly crashed into a school bus approaching from the opposite direction. Lieutenant Buchanan was treated at the scene and transported to the hospital. Despite these efforts, he expired.

⌒

September 21, 2006 – 1930hrs
John A. Beyer, Firefighter
Age 38, Volunteer
Wilson Volunteer Fire Company, New York

Firefighter Beyer participated in a 2-hour auto extrication training exercise during the evening of September 21, 2006. He went home after training and collapsed of a heart attack approximately 2 hours after the completion of training. He was transported to a local hospital and successfully resuscitated. He was transferred to a regional hospital where his condition deteriorated. Firefighter Beyer died on September 22, 2006.

⌒

September 22, 2006 – 1030hrs
Ronald Phillip Allen, Jr., Lieutenant
Age 36, Volunteer
Tar Heel Fire Department, North Carolina

Lieutenant Allen was marking the location of fire hydrants by spray painting arrows on the roadway pavement. Lieutenant Allen used a brush truck to block the lane in which he was working. The hazard lights of the vehicle were flashing.

The brush truck was struck from behind by another vehicle. The truck was propelled forward and struck Lieutenant Allen. He was trapped under the vehicle and was pronounced dead at the scene.

⌒

September 27, 2006 – 1352hrs
Edward John Jenik, Executive Captain
Age 53, Career
Highland Heights Fire Department, Ohio

Executive Captain Jenik went to his home for lunch on September 27, 2006. He was leaving his house to return to work when he suffered a heart attack. A neighbor found him unconscious behind the wheel of his fire department vehicle in the roadway and called 9-1-1.

Executive Captain Jenik was transported to the hospital by ambulance but was pronounced dead approximately 30 minutes after his arrival.

October 10, 2006 – 0230hrs
Allan M. Roberts, Firefighter
Age 40, Career
Baltimore City Fire Department, Maryland

Firefighter Roberts was assigned to Engine 41 for the night shift of October 9, 2006. Engine 41 and other fire companies were dispatched to a report of a residential structure fire at 0222hrs on October 10, 2006.

Engine 41 was the first to arrive on the scene and found a working fire. Two occupants of the home had jumped from an upper-story window, and there were reports of other trapped occupants. Firefighter Roberts and two other firefighters entered the structure to perform a search and fight the fire.

At approximately 0300hrs, Firefighter Roberts and other firefighters were advancing a charged handline up the interior stairs to the second floor when an extremely rapid buildup of heat occurred. Firefighters were forced to abandon the fire fight, the front door of the residence somehow became closed, and firefighters were trapped in the building. Other firefighters were able to rescue Firefighter Allen, and he was removed from the structure.

Somehow during the attempt to escape the structure, Firefighter Allen's helmet and facepiece were removed. Firefighter Allen was transported to the hospital but was not revived.

The cause of death was listed as smoke inhalation. Firefighter Allen's carboxyhemoglobin level at autopsy was 26 percent. The fire was caused by an electrical short in a hot water heater.

October 16, 2006 – 1330hrs
John A. Stura, Firefighter
Age 78, Volunteer
North Belle Vernon Fire Department, Pennsylvania

Firefighter Stura was on his way to the bank to deposit the proceeds of a fire department bingo fundraiser. Firefighter Stura was walking near the bank parking lot when he was struck by a vehicle.

After striking Firefighter Stura, the driver of the vehicle accidentally depressed the accelerator rather than the brake. The vehicle continued through the parking lot until it came into contact with a raised landscaped area. Firefighter Stura was trapped under the vehicle. He was removed by firefighters and transported to the hospital, but he did not survive. The cause of death was listed as blunt force trauma of the chest.

October 19, 2006 – 1200hrs
Raleigh Eugene England II, Chaplain/Deputy Chief
Age 59, Volunteer
Beaver Volunteer Fire Department, West Virginia

Deputy Chief England participated in a weekly work detail at his fire station. While working, he complained of not feeling well. Shortly thereafter, he departed for lunch. During the lunch break, Deputy Chief England suffered a CVA (stroke). He died on October 28, 2006.

October 26, 2006 – 0730hrs
Mark Allen Loutzenhiser, Captain
Age 43, Wildland Full-Time
USDA Forest Service - San Bernardino National Forest, California

Jess Edward McLean, Engineer
Age 27, Wildland Full-Time
USDA Forest Service - San Bernardino National Forest, California

Pablo Cerda, Firefighter
Age 24, Wildland Part-Time
USDA Forest Service - San Bernardino National Forest, California

Daniel Kurtis Hoover-Najera, Firefighter
Age 20, Wildland Part-Time
USDA Forest Service - San Bernardino National Forest, California

Jason Robert McKay, Firefighter
Age 27, Wildland Full-Time
USDA Forest Service - San Bernardino National Forest, California

Captain Loutzenhiser, Engineer McLean, Firefighter Cerda, Firefighter Hoover-Najera, and Firefighter McKay were assigned as the crew of Engine 57.

At approximately 0111hrs on October 26, 2006, a fire was intentionally set at the bottom of a slope near the town of Cabazon, California. The fire quickly spread uphill toward a highway and the rural community of Twin Pines. The fire spread into an area where firefighting responsibilities are provided by the United States Department of Agriculture Forest Service (USDAFS). At 0130hrs, the local IC requested five USDAFS engine companies. By 0307hrs, the fire had grown to over 500 acres.

By 0402hrs, the five engine companies, including Engine 57, had arrived at the Command Post. The companies were assigned to structure protection and evacuation duties. Engine 57 was assigned to protect a residential structure and discussed plans with a command officer at approximately 0620hrs. The crew deployed hoselines from their apparatus and from a portable pump drawing water from a pool.

Shortly after 0700hrs, the fire progressed rapidly and overcame the position held by Engine 57. Winds during this period exceeded 50 mph, and the flame front was reported to be 90 feet tall. Other fire crews responded to the scene and provided treatment to the crew of Engine 57.

Engineer McLean, Firefighter McKay, and Firefighter Hoover-Najera were pronounced dead at the scene; Captain Loutzenhiser was transported to the hospital but died. Firefighter Cerda was transported to the hospital and died as a result of his injures on October 31, 2006. All five firefighters died as the result of burns.

November 1, 2006 – 0415hrs
Gregory Allen "Greg" Cloud, Firefighter
Age 32, Volunteer
Kent Volunteer Fire Company, Indiana

Firefighter Cloud and the members of his fire department were dispatched to provide mutual-aid at a structure fire in a nearby town. Firefighter Cloud was one of the first members of his department to arrive at the scene of the working fire in a very large residence. The IC ordered Firefighter Cloud and another Kent firefighter to don SCBAs and prepare to mount an interior attack on the fire.

Fire had started in a gazebo behind the house and spread to the rear of the house. Firefighter Cloud and another firefighter entered the structure and proceeded to the second floor. The firefighters encountered extreme levels of heat and smoke on the second floor and decided to return to the ground level. As the firefighters left the building, fire progressed rapidly.

Once outside, firefighters discovered that Firefighter Cloud had not escaped. Firefighters fought their way back into the structure and located Firefighter Cloud on the second floor. He was located by the sound of his PASS device. Firefighter Cloud was obviously deceased; he had sustained major burns, and his SCBA facepiece had melted.

At autopsy, Firefighter Cloud's carboxyhemoglobin level was 38 percent.

November 3, 2006 – 1845hrs
Joseph S. Pagano, Captain
Age 52, Career
Middletown Fire Department, Connecticut

Captain Pagano suffered a heart attack while on duty and working at his desk in the fire station. Other on-duty firefighters provided medical assistance, but their efforts were not successful in reviving Captain Pagano.

November 12, 2006 – 1138hrs
Kyle William Clarence Weisbrich, Firefighter
Age 22, Volunteer
Melrose Fire Department, Minnesota

Firefighter Weisbrich was responding to the fire station on his motorcycle after his department was dispatched to a motor vehicle crash. During the response, Firefighter Weisbrich failed to stop for a stop sign, entered the intersection, and struck a vehicle that was in the intersection.

Firefighters arriving on the scene provided medical care to Firefighter Weisbrich. He was transported via ambulance to the hospital, but was not revived. The cause of death was listed as traumatic head injuries. The speed of Firefighter Weisbrich's motorcycle was cited in the law enforcement crash report.

November 14, 2006 – 0106hrs
Robert Gerald Whittaker, Firefighter
Age 55, Volunteer
Marshallberg Voluntary Fire Department, North Carolina

Firefighter Whittaker was the driver of a heavy rescue apparatus responding to the scene of an arson-caused mutual-aid structure fire. Firefighter Whittaker suffered a heart attack, the apparatus left the roadway, crossed a ditch, and struck a tree.

Other firefighters riding in the apparatus, who were all wearing their seatbelts at the time of the crash, were uninjured and provided emergency medical assistance. Firefighter Whittaker was transported to the hospital but was not revived.

⌒

November 16, 2006 – 0654hrs
Michael Timothy Browne, Firefighter/EMT-B
Age 25, Volunteer
Acme-Delco-Riegelwood Fire and Rescue, North Carolina

Firefighter Browne was leaving his home in response to a fire department page for severe weather. As he left home, the area was struck with a tornado that overtook him.

Firefighter Browne was severely injured and died at the scene. His father and his stepmother also were killed. Firefighter Browne's 3-year-old daughter survived.

A total of eight people died as a result of the tornado, 35 homes were damaged or destroyed, and approximately 100 people were left homeless in the category F3 storm.

⌒

November 23, 2006 – 2000hrs
Steven M. Solomon, Firefighter
Age 33, Career
Atlanta Fire-Rescue Department, Georgia

Firefighter Solomon was working a 24-hour shift on Thanksgiving Day. Shortly after 2000hrs, Atlanta Fire-Rescue dispatched a full first-alarm assignment for a reported fire in an abandoned house. On arrival, companies encountered heavy smoke showing from a boarded-up single-story brick structure. As other crews removed plywood window coverings and forced entry through the front door, the crew of Engine 16 prepared to advance a 1-3/4-inch attack line into the house.

Firefighter Solomon was on the nozzle as the line was advanced inside. The attack team immediately encountered high temperature and zero-visibility conditions. Within seconds after they entered, the battalion chief arrived, assumed command, and ordered the companies to operate in a defensive strategy. Before the line could be backed out, the interior became enveloped in flames and the three firefighters from Engine 16 lost track of each other. Two of the firefighters managed to escape through the front door.

Firefighters outside saw the silhouette of a firefighter, enveloped in flames, running past the front door and moving toward the rear of the house. The fire was quickly knocked down and crews made entry from both

continued on next page

the front and rear to conduct a search. Firefighter Solomon was located almost immediately by a member who was using a thermal imaging camera, and several firefighters quickly removed him from the dwelling. He was unconscious and critically burned. When he was found, Firefighter Solomon had removed his helmet, hood, and SCBA facepiece. One boot also was missing.

Although he received immediate treatment from firefighter/paramedics on the scene and was transported within minutes to a level-one trauma center and regional burn unit, Firefighter Solomon died 6 days later without regaining consciousness.

⤸

November 26, 2006 – 1348hrs
Hector "Sandy" McClune, Firefighter
Age 76, Volunteer
Millerton Fire Department, New York

Firefighter McClune and approximately 25 firefighters were battling a wildland fire at a local school. Firefighter McClune suddenly collapsed due to a heart attack. Firefighters provided medical treatment and he was transported by ambulance to the hospital.

Despite the efforts of firefighters and other responders, Firefighter McClune died at the hospital.

⤸

November 28, 2006 – 0810hrs
Thomas J. Van Liew, Acting Lieutenant
Age 52, Career
New Brunswick Fire Department, New Jersey

Acting Lieutenant Van Liew had been on duty for a short time and working the floor watch. He was found unresponsive by other firefighters.

Paramedics were summoned, and firefighters provided treatment to Acting Lieutenant Van Liew. He was transported to the hospital but died as the result of a heart attack.

⤸

November 30, 2006 – 0200hrs
Jeffrey Scott Hollingsworth, Deputy Chief
Age 38, Volunteer
Clement Volunteer Fire Department, North Carolina

Deputy Chief Hollingsworth and other members of his fire department responded to a structure fire and a small wildland fire. The incident was concluded at approximately 0207hrs.

At approximately 0700hrs, Deputy Chief Hollingsworth complained of chest pain and was evaluated by a physician. He was sent home, but returned to the hospital later that day still feeling ill. He was admitted but died later that day as the result of a heart attack.

⤸

December 1, 2006 – 0257hrs
Leo Howard Soderquist, Firefighter
Age 64, Volunteer
Axtell Fire and Rescue, Nebraska

Firefighter Soderquist and the members of his fire department responded to the scene of a structure fire in a residence. While on the scene, Firefighter Soderquist pulled and set up handlines, set up a blower fan, did salvage, and overhaul, and helped with the return of equipment to the apparatus. The fire, started by an electrical problem, was primarily confined to the attic of the home.

Firefighters returned to the fire station. Firefighter Soderquist was sitting on the back step of a fire apparatus talking to other firefighters when he suddenly collapsed. He was treated by firefighters in the fire station and transported to the hospital. His death was caused by a heart attack.

ᔇ

December 1, 2006 – 1745hrs
Kent Furman Long, Firefighter II
Age 44, Career
Charlotte Fire Department, North Carolina

Firefighter Long was on duty in his regularly assigned fire station. The day had been unremarkable with responses to incidents and a move-up to another fire station that was concluded at approximately 1545hrs. Firefighter Long's last emergency response was an emergency medical incident at 1642hrs.

At approximately 1730hrs, Firefighter Long received permission from his company officer to exercise in a parking lot next door to the fire station.

A short time after his departure, a passerby knocked on the door of the fire station and advised that someone was down on the grass in front of the fire station. Firefighters responded outside and found the person down was Firefighter Long. CPR was initiated, and paramedic-level emergency medical care was provided.

Firefighter Long was transported to the hospital where efforts to revive him were continued to no avail. The cause of death was listed as a heart attack.

ᔇ

December 7, 2006 – Time Unknown
Thomas Joseph Hays, Firefighter
Age 25, Volunteer
Lower Merion Fire Department / Narberth Fire Company, Pennsylvania

Firefighter Hays responded to a commercial structure fire at 2030hrs on December 7, 2006. Firefighter Hays helped open up the walls of the structure to provide access for firefighters. The incident was concluded by approximately 2200hrs and Firefighter Hays returned to his home.

Concerned that he did not respond to fire incidents that occurred over the early morning hours of December 8, 2006, firefighters went to Firefighter Hays' home to check on his welfare. They discovered Firefighter Hays deceased in his bed.

The death was attributed to epilepsy.

⤳

December 9, 2006 – 1620hrs
Edward DeWitt Wilburn, President
Age 64, Volunteer
Deep Creek Volunteer Fire Department, Maryland

President Wilburn was responding in his personal vehicle to a report of a working structure fire. During the response, President Wilburn suffered a heart attack. He stopped his vehicle in the roadway. Arriving EMS responders found President Wilburn slumped over the wheel. Responders tapped on the window and the vehicle lurched forward, traveling approximately 100 yards on a grassy shoulder before reaching and cresting a steep embankment. The car came to rest upon reaching the other side of the embankment.

Firefighters removed President Wilburn from the vehicle and began CPR. He was transported to the hospital where he was pronounced dead.

⤳

December 28, 2006 – 1300hrs
Cecil Tackett, Jr., Firefighter
Age 28, Volunteer
Flat Gap Fire Department, Kentucky

Firefighter Tackett helped set up for a community function at his fire station. During the function, Firefighter Tackett went home. When it was time for the function to be over, Firefighter Tackett drove his personal vehicle back to the station. He had been assigned to clean up after the event and put the fire apparatus back into their bays.

As he drove to the station, he was involved in a vehicle crash and was killed.

⤳

December 29, 2006 – 0557hrs
Stephen Jones, Firefighter
Age 56, Volunteer
Barnstead Fire & Rescue, Inc., New Hampshire

Firefighter Jones had been assigned by his fire chief to bring a piece of fire apparatus to a maintenance facility in another town. As he drove between the town's fire stations, he began to experience chest pains and called for help.

Firefighter Jones was transported to the hospital but died as a result of a heart attack.

⤳

December 30, 2006 – 0913hrs
Philip Wayne Townsend, Firefighter
Age 31, Career
Denison Fire Department, Texas

Firefighter Townsend and the members of his fire department were dispatched to a commercial structure fire. Upon their arrival, firefighters found a working attic fire in a single-story strip center.

Firefighters attempted an interior attack on the fire but were driven from the building by heavy smoke and fire conditions. A defensive mode was declared, and firefighters left the interior of the building.

Firefighter Townsend and the fire chief were standing approximately 2 feet from the building when a structural collapse occurred. Both firefighters were trapped in the debris. Firefighters and bystanders immediately began to remove debris to access the trapped firefighters.

The fire chief was removed from the debris within approximately 9 minutes of the collapse. He was transported to the hospital and treated for a separated shoulder and a head contusion.

Firefighter Townsend was extricated after approximately 11 minutes. He was transported to the hospital but he was pronounced dead. The cause of death was listed as blunt force trauma.

Incidents Prior To 2006

October 21, 1990 – 1406hrs
George Malcolm Jackson, Firefighter
Age 51, Career
Camden Fire Department, New Jersey

Firefighter Jackson was assigned to an engine company. On October 21, 1990, his engine was dispatched alone to a report of a stove fire. Upon their arrival on the scene, firefighters found medium smoke showing from the rear of an occupied two-story brick rowhouse. The company officer called for the balance of a box assignment.

Firefighter Jackson and his crew advanced a handline into the structure and conducted a search of the first floor. When ladder company firefighters arrived and entered the building, Firefighter Jackson joined a ladder crew and ascended to the second floor of the house.

The fire in the first floor kitchen was proving difficult to control, and the pump operator was having difficulty opening a fire hydrant in front of the house. The pump operator radioed that crews should back out due to a water supply interruption. The fire in the kitchen advanced rapidly.

Firefighters, including Firefighter Jackson, had completed their search of the second floor, but were unable to exit the structure due to heat and fire conditions on the stairs. The three firefighters retreated to the second floor. The ladder company firefighters were able to escape the building by diving head-first through windows; Firefighter Jackson was overcome before he could escape.

Later-arriving fire companies established an alternate water supply and darkened down the fire on the second floor. Firefighters accessed the second floor over ground ladders and found Firefighter Jackson just inside a window. Firefighter Jackson was removed from the building, suffering from critical hand and face burns.

Firefighter Jackson was transported to the hospital by ambulance. Firefighter Jackson remained in a coma until his death on March 7, 2006, at age 67.

December 29, 1995 – 0650hrs
Donald Joseph Herbert, Firefighter
Age 34, Career
Buffalo Fire Department, New York

Firefighter Herbert was fighting a fire in a three-story residential structure. He was inside the structure when a roof collapse occurred. Firefighter Herbert was pinned by the collapse. He was removed from the structure in cardiac arrest and transported to the hospital.

Firefighter Herbert was revived, but lapsed into a coma. He regained consciousness occasionally over a 9-year period. Firefighter Herbert died as a result of his injuries on February 21, 2006, at the age of 44.

August 12, 2002 – 1945hrs
Charles Eugene Kessel, Firefighter
Age 52, Volunteer
Maysville Volunteer Fire Department, West Virginia

Firefighter Kessel was transporting fire department fundraising equipment in his pickup truck from a local fairground to the fire station. Some of the items fell off the back of the truck. Firefighter Kessel stopped to pick up the items.

While he was away from the vehicle, Firefighter Kessel realized that the vehicle was rolling away. He attempted to stop the vehicle but did not make it fully into the vehicle. Firefighter Kessel was dragged several feet before the vehicle hit a culvert and a road sign. Firefighter Kessel's head was caught between the pickup truck and the road sign and he received serious facial and head injuries.

Firefighter Kessel never fully recovered from his injuries. He died on October 16, 2006, as a result of complications of his injury.

APPENDIX B

FIREFIGHTER FATALITY INCLUSION CRITERIA – NATIONAL FIRE SERVICE ORGANIZATIONS

The National Fire Protection Association (NFPA), the National Fallen Firefighters Foundation (NFFF), the United States Fire Administration (USFA), and other organizations collect information on firefighter fatalities in the United States. Each organization uses a slightly different set of inclusion criteria that are based at least in part on the purposes of the information collection for each organization and data consistency.

As a result of these differing inclusion criteria, statistics about firefighter fatalities may be provided by each organization that do not coincide with one another. This section will explain the inclusion criteria for each organization and provide information about these differences.

The USFA includes firefighters in this report who die while on duty, become ill while on duty and die later, and firefighters who die within 24 hours of an emergency response or training, regardless of whether the firefighter complained of illness while on duty. The USFA counts firefighter deaths that occur in the 50 States, the District of Columbia, and United States protectorates such as Puerto Rico and Guam. Detailed inclusion criteria for this report appear starting on page 98 of this report.

For 2006, the USFA reported 106 onduty firefighter fatalities.

Firefighter Fatalities for Incidents Occurring in 2006

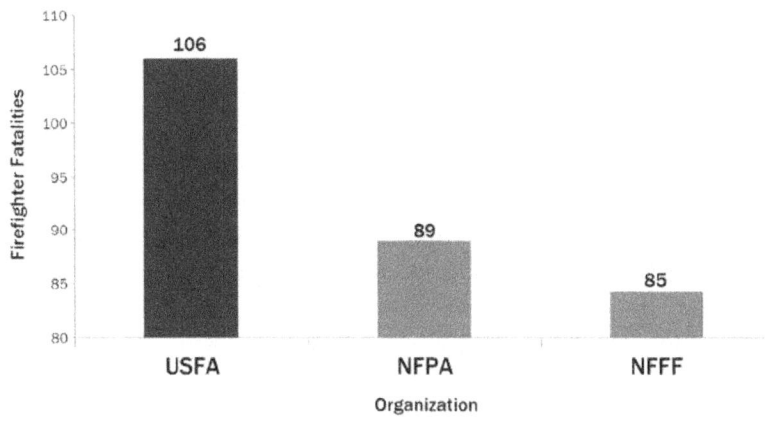

INCLUSION CRITERIA FOR NFPA'S ANNUAL FIREFIGHTER FATALITY STUDY

Introduction

Each year, the National Fire Protection Association (NFPA) collects data on all firefighter fatalities in the United States that resulted from injuries or illnesses that occurred while the victims were on duty. The purpose of the study is to analyze trends in the types of illnesses and injuries resulting in death that occur while firefighters are on the job. This annual census of firefighter fatalities in its current format dates back to 1977. (Between 1974 and 1976, NFPA published a study of onduty firefighter fatalities that was not as comprehensive.)

What is a Firefighter?

For the purpose of the NFPA study, the term firefighter covers all uniformed members of organized fire departments, whether career, volunteer, combination, or contract; full-time public service officers acting as firefighters; State and Federal government fire service personnel; temporary fire suppression personnel operating under official auspices of one of the above; and privately employed firefighters including trained members of industrial or institutional fire brigades, whether full- or part-time.

Under this definition, the study includes, besides uniformed members of local career and volunteer fire departments, those seasonal and full-time employees of State and Federal agencies who have fire suppression responsibilities as part of their job description, prison inmates serving on firefighting crews, military personnel performing assigned fire suppression activities, civilian firefighters working at military installations and members of industrial fire brigades. Impressed civilians also would be included if called on by the officer in charge of the incident to carry out specific duties. The NFPA study includes fatalities that occur in the 50 States and the District of Columbia.

What does 'on duty' mean?

The term on duty refers to being at the scene of an alarm, whether a fire or non-fire incident; being en route while responding to or returning from an alarm; performing other assigned duties such as training, maintenance, public education, inspection, investigations, court testimony, or fundraising; and being on call, under orders or on standby duty other than at home or at the individual's place of business. Fatalities that occur at a firefighter's home may be counted, if the actions of the firefighter at the time of injury involved firefighting or rescue.

Onduty fatalities include any injury sustained in the line of duty that proves fatal, any illness that was incurred as a result of actions while on duty that proves fatal, and fatal mishaps involving nonemergency occupational hazards that occur while on duty. The types of injuries included in the first category are mainly those that occur at an incident scene, in training, or in accidents while responding to or returning from alarms. Illnesses (including heart attacks) are included when the exposure or onset of symptoms are tied to a specific incident of onduty activity. Those symptoms must have been in evidence while the victim was on duty for the fatality to be included in the study.

Fatal injuries and illnesses are included even in cases where death is considerably delayed. When the onset of the condition and the death occur in different years, the incident is counted in the year of the condition's onset. Medical documentation specifically tying the death to the specific injury is required for inclusion of these cases in the study.

Categories not included in the study

The NFPA study does not include members of fire department auxiliaries; nonuniformed employees of fire departments; emergency medical technicians who are not also firefighters; chaplains; or civilian dispatchers. The study also does not include suicides as onduty fatalities, even when the suicide occurs on fire department property.

The NFPA recognizes that a comprehensive study of firefighter onduty fatalities would include chronic illnesses (such as cardiovascular disease and certain cancers) that prove fatal and that arose from occupational factors. In practice, there is as yet no mechanism for identifying onduty fatalities that are due to illnesses that develop over long periods of time. This creates an incomplete picture when comparing occupational illnesses to other factors as causes of firefighter deaths. This is recognized as a gap the size of which cannot be identified at this time because of the limitations in tracking the exposure of firefighters to toxic environments and substances and the potential long-term effects of such exposures.

2006 Experience

In 2006, a total of 89 onduty firefighter deaths occurred in the United States, according to the NFPA inclusion criteria.

APPENDIX C

HOMETOWN HEROES SURVIVORS BENEFIT ACT OF 2003

PUBLIC LAW 108–182—DEC. 15, 2003   117 STAT. 2649

Public Law 108–182
108th Congress

An Act

To ensure that a public safety officer who suffers a fatal heart attack or stroke while on duty shall be presumed to have died in the line of duty for purposes of public safety officer survivor benefits.

Dec. 15, 2003

[S. 459]

Be it enacted by the Senate and House of Representatives of the United States of America in Congress assembled,

Hometown
Heroes Survivors
Benefits Act of
2003.
42 USC 3711
note.

SECTION 1. SHORT TITLE.

This Act may be cited as the "Hometown Heroes Survivors Benefits Act of 2003".

SEC. 2. FATAL HEART ATTACK OR STROKE ON DUTY PRESUMED TO BE DEATH IN LINE OF DUTY FOR PURPOSES OF PUBLIC SAFETY OFFICER SURVIVOR BENEFITS.

Section 1201 of the Omnibus Crime Control and Safe Streets Act of 1968 (42 U.S.C. 3796) is amended by adding at the end the following:

"(k) For purposes of this section, if a public safety officer dies as the direct and proximate result of a heart attack or stroke, that officer shall be presumed to have died as the direct and proximate result of a personal injury sustained in the line of duty, if—

"(1) that officer, while on duty—

"(A) engaged in a situation, and such engagement involved nonroutine stressful or strenuous physical law enforcement, fire suppression, rescue, hazardous material response, emergency medical services, prison security, disaster relief, or other emergency response activity; or

"(B) participated in a training exercise, and such participation involved nonroutine stressful or strenuous physical activity;

"(2) that officer died as a result of a heart attack or stroke suffered—

"(A) while engaging or participating as described under paragraph (1);

"(B) while still on that duty after so engaging or participating; or

"(C) not later than 24 hours after so engaging or participating; and

"(3) such presumption is not overcome by competent medical evidence to the contrary.

continued on next page

Appendix C (continued)

117 STAT. 2650   PUBLIC LAW 108–182—DEC. 15, 2003

"(l) For purposes of subsection (k), 'nonroutine stressful or strenuous physical' excludes actions of a clerical, administrative, or nonmanual nature.".

Approved December 15, 2003.

LEGISLATIVE HISTORY—S. 459:

CONGRESSIONAL RECORD, Vol. 149 (2003):
   May 15, considerd and passed Senate.
   Nov. 21, considered and passed House, amended.
   Nov. 25, Senate concurred in House amendment.
                              ○

Appendix D

National Fallen Firefighters Foundation

In 1997, fire service leaders formulated new criteria to determine eligibility for inclusion on the National Fallen Firefighter Memorial. Line-of-duty deaths shall be determined by the following standards:

1. (a) Deaths of firefighters meeting the Department of Justice's Public Safety Officers' Benefits (PSOB) program guidelines, and those cases that appear to meet these guidelines whether or not PSOB staff has adjudicated the specific case prior to the annual National Fallen Firefighters Memorial Service; and

   (b) Deaths of firefighters from injuries, heart attacks or illnesses documented to show a direct link to a specific emergency incident or department-mandated training activity.

2. While PSOB guidelines cover only public safety officers, the Foundation's criteria also include contract firefighters and firefighters employed by a private company, such as those in an industrial brigade, provided that the deaths meet the standards listed above.

3. Some specific cases will be excluded from consideration, such as deaths attributable to suicide, alcohol or substance abuse, or other gross abuses as specified in the PSOB guidelines.

The National Fallen Firefighters Memorial was built in 1981 in Emmitsburg, MD. The names listed there begin with those firefighters who died in the line-of-duty that year. The United States Congress created the National Fallen Firefighters Foundation to lead a nationwide effort to remember America's fallen firefighters. Since 1992, the tax-exempt, nonprofit Foundation has developed and expanded programs to honor our fallen fire heroes and assist their families and coworkers by providing them with resources to rebuild their lives. Since 1997, the Foundation has managed the National Memorial Service held each October to honor the firefighters who died in the line-of-duty the previous year.

At the October 2007 Memorial Weekend, the Foundation will be honoring 91 firefighters who died in the line-of-duty. Of those 91 being honored, 85 died in 2006 as the result of incidents that occurred in 2006, 2 firefighters died in 2006 as the result of incidents that occurred in previous years, and 4 others died in previous years as the result of incidents that occurred in previous years. The following section is a listing of the firefighters who will be honored by the Foundation in October of 2007.

Firefighter deaths that occurred in 2006 as the result of an incident that occurred in 2006:

Ronald Phillip Allen, Jr., Lieutenant
Allan H. Anderson, Jr., Firefighter/Rescue Diver
Kevin Anthony Apuzzio, Foreman
Roger W. Armstrong, Firefighter
John A. Beyer, Firefighter
Joseph Louis Bilka, Firefighter
Jeffrey A. Bowman, Lieutenant
Howard John Carpluk, Jr., Lieutenant
Pablo Cerda, Firefighter
Tracy Champion, Firefighter
Gregory Allen "Greg" Cloud, Firefighter
William Leroy Craddock II, Firefighter III
Michael Lynn Davenport, Inmate Firefighter
Michael James Day, Deputy Assistant Chief
Robert Wallace "Bobby" Edwards, Firefighter
Larry Wayne Fanning, Firefighter
Michael J. Greene, Firefighter
Dennis K. "Denny" Hayes, Firefighter
Thomas Joseph Hays, Firefighter
Jeffrey Scott Hollingsworth, Deputy Chief
Daniel Kurtis Hoover-Najera, Firefighter
John Destry Horton, Firefighter/Paramedic
Jeffrey Dean Irwin, Firefighter
Terry Wayne Jacobs, Pilot
Jason Allen Johnson, Firefighter
Dustin Keith Jones, Firefighter
Kelly Michael Kincaid, Lieutenant
Spencer Stanley Koyle, Division Supervisor
Thomas James "Emmett" Kuehl, Firefighter/EMT
Alan Dale Leake, Captain
Michael Gene Lewis, Firefighter
Kent Furman Long, Firefighter II
Richard O. Longoria, Firefighter II/Paramedic
Mark Allen Loutzenhiser, Captain
Dana Lynn MacCrimmon, Firefighter
Gerald A. Machajewski, Firefighter
Edward Joseph Marbet, Firefighter
Hector "Sandy" McClune, Firefighter
Lloyd B. McCulloch, Captain
Jason Robert McKay, Firefighter
Robert Eugene McLaughlin, Captain
Jess Edward McLean, Engineer
James Wilson McMorries, Jr., Firefighter
John Paul "JP" Memory II, Firefighter

Errett Wayne Miller, Assistant Chief
Paul Raymond Montavon Sr., Lieutenant
Richard Patrick "Rich" Montoya, Lieutenant
David Louis Moore II, Assistant Chief
Vincent R. Neglia, Acting Captain
Eric James Olson, Firefighter
David Robert Packard, Lieutenant
Joseph S. Pagano, Captain
Lillian M. Patten, Firefighter/Lookout
Willie Macon Price, Firefighter
Jose Luis Ramirez Jr., Firefighter/EMT
Michael C. Reilly, Firefighter
Wilbur A. Ritter, Fire Police Officer
Allan M. Roberts, Firefighter
Amy Lynn Schnearle-Pennywitt, Firefighter
Robert John Schnibbe, Jr., Battalion Chief
Russell Blanton Schwantes,
    Fire Apparatus Operator
Steven A. Smith, Deputy Chief
David E. Smith, Fire Police Officer/
    Fire Commissioner
Leo Howard Soderquist, Firefighter
Steven M. Solomon, Firefighter
Mark Allen Stanley, Forestry Technician
Robert Paul Stone, Battalion Chief
Quinn Russell Stone, Pilot
Richard George "Dick" Sullivan, Fire Chief
Garry Tanner, Assistant Chief
Ollie Phil Tate, Firefighter
Harold Vernon Taylor, Assistant Chief
Phillip Wayne Townsend, Firefighter
Lee William Walters, Assistant Chief
Richard Washington Jr., Firefighter II
Kyle William Clarence Weisbrich,
    Probationary Firefighter
John Robert "Bobby" Westervelt, President
Robert Gerald Whittaker, Firefighter
Edward DeWitt Wilburn, Firefighter
George Elvin "Sandy" Willett Jr., Pilot
Robert "Ockie" Wisting, Firefighter
Arnold W. "Arnie" Wolff, Lieutenant
Rose Ann Woodbridge, Firefighter
Wayne Edward "Butch" Yarborough, Lieutenant
Monica Lee Zajanc, Firefighter

Firefighter deaths that occurred in 2006 from incidents in previous years:

George Malcolm Jackson, Firefighter

Donald Joseph Herbert, Firefighter

Firefighter deaths that occurred in previous years:

Joseph Ezzo, Captain
Tommy Kidd, Firefighter

Dennis R. Lemery, Firefighter
Woodrow W. Poland, Sr., President

Photo by Bill Green

This photo was taken at the National Fallen Firefighters Memorial Service in October 2006.

www.FireFighterNearMiss.com • *www.FireHero.org* • *www.EveryoneGoesHome.com*

www.ingramcontent.com/pod-product-compliance
Lightning Source LLC
Chambersburg PA
CBHW081137170526

45165CB00008B/2703